Mathematical Research

Geobild '89

edited by
A. Hübler · W. Nagel
B. D. Ripley · G. Werner

Volume 51

AKADEMIE-VERLAG BERLIN

In this series original contributions of mathematical research in all fields are contained, such as

— research monographs
— collections of papers to a single topic
— reports on congresses of exceptional interest for mathematical research.

This series is aimed at promoting quick information and communication between mathematicians of the various special branches.

In diese Reihe werden Originalbeiträge zu allen Gebieten der mathematischen Forschung aufgenommen wie

— Forschungsmonographien
— Sammlungen von Arbeiten zu einem speziellen Thema
— Berichte von Tagungen, die für die mathematische Forschung besonders aktuell sind.

Die Reihe soll die schnelle Information und gute Kommunikation zwischen den Mathematikern der verschiedenen Fachgebiete fördern.

Manuscripts
in English and German comprising at least 100 pages and not more than 500 pages can be admitted to this series. With respect to a quick publication the manuscripts are reproduced photomechanically. Authors who are interested in this series please turn directly to the 'Akademie-Verlag'. Here you will get more detailed information about the form of the manuscripts and the modalities of publication.

Manuskripte
in englischer und deutscher Sprache, die mindestens 100 Seiten und nicht mehr als 500 Seiten umfassen, können in diese Reihe aufgenommen werden. Im Interesse einer schnellen Publikation werden die Manuskripte auf fotomechanischem Weg reproduziert. Autoren, die an der Veröffentlichung entsprechender Arbeiten in dieser Reihe interessiert sind, wenden sich bitte direkt an den Akademie-Verlag. Sie erhalten dort genauere Informationen über die Gestaltung der Manuskripte und die Modalitäten der Veröffentlichung.

Geobild '89

Mathematical Research · Mathematische Forschung

Wissenschaftliche Beiträge
herausgegeben von der
Akademie der Wissenschaften der DDR
Karl-Weierstraß-Institut für Mathematik

Band 51

Geobild '89

Geobild '89

Proceedings of the 4th Workshop on Geometrical Problems of Image Processing held in Georgenthal (GDR), March 13 – 17, 1989

edited by

Albrecht Hübler

Werner Nagel

Brian David Ripley

Günter Werner

Akademie-Verlag Berlin 1989

Herausgeber:

Dr. rer. nat. Albrecht Hübler
Dr. rer. nat. Werner Nagel
Dr. rer. nat. Günter Werner

Friedrich-Schiller-Universität Jena
Sektion Mathematik

Prof. Dr. Brian David Ripley
Department of Mathematics
University of Strathclyde

Die Titel dieser Schriftenreihe werden vom Originalmanuskript der Autoren reproduziert.

ISBN 3-05-500677-1
ISSN 0138-3019

Erschienen im Akademie-Verlag Berlin,
Leipziger Str. 3 - 4, Berlin, DDR-1086

Lizenznummer: 202·100/411/89
Printed in the German Democratic Republic
Gesamtherstellung: VEB Kongreß- und Werbedruck,
Oberlungwitz, DDR-9273
Lektor: Dr. Reinhard Höppner
LSV 1085
Bestellnummer: 763 790 6 (2182/51)

02800

PREFACE

At the present time there are a lot of good algorithms and effective tools for Image Processing and Analysis. Several of them are based upon solid theoretical foundations. But there is an obvious need for a further involvement of Mathematics. The theory should provide clear and consistent formulations of problems, definitions and notations as well as a systematic approach and, of course, new ideas! We - and, we think, the authors too - find this a provoking challenge.
It is one intention of the "Geobild"-workshops to reflect the state and the current trends of the geometric and stochastic foundations of Image Processing and Analysis. The emphasis is on the fields of Digital Geometry - Computational Geometry - Integral Geometry - Stochastic Geometry - Stereology. According to the mind of the workshop several tutorial and review papers are included.
Digital Geometry, as it has developed in the last 20 years, investigates the interconnections between traditional continuous geometry and its image in discrete grids with respect to given digitization mappings. Important problems are the characterization of digitized objects like straight lines, circles, convex sets etc. and their parameters, as well as the provision of practicable algorithmic solutions of corresponding computational problems. For several years there has been a growing interest in the study of the discrete image structures themselves independent of a continuous geometry. The main aims of current activity in this area are to understand and to describe the discrete topology and geometry inherent in the supporting domain of discrete images. The papers in this volume which refer to this subject should be understood as first attempts in this direction.
Computational Geometry as we understand it, was introduced ten years ago. Current trends are given in an incomplete list of typical geometrical problems: convexity, intersection, visibility, point location, contour, shortest path motion, partition and computation of parameters such as distances, perimeter. Many new solutions to computational problems are now available. Design principles for algorithms such as the sweep-line paradigm, divide and conquer, geometric transformations, belong in the toolbox of theorists and practitioners working in this area. Classical data structures like Voronoi-diagrams were used successfully and new structures such as segment-, interval-, and priority search trees have been invented. In Integral Geometry some progress has been made with respect to translative integral formulae, i.e. here, as opposed to the classical theory, only translations but not rotations are taken into account. This meets the practical need of an investigation of the anisotropy of structures and

of parallel sectioning in Stereology. Furthermore, new techniques (e.g. confocal microscopes) allow the observation of the joint distribution of profiles on several section planes through a specimen. This must be exploited by new integral geometric and stereological formulae for second order moments and quantities of geometrical structures. Such quantities are necessary in order to make headway with regard to the problems of analyzing the spatial arrangement of particles. In Stochastic Geometry the theory of random sets is developing further and the store of stochastic models for real structures is being enriched.

The theory of Fuzzy Sets is becoming increasingly important in the treatment of grey tone images. Therefore, a longer tutorial paper on this topic is included. In the literature this approach is also introduced under the name grey-scale morphology.

For the convenience of the reader, the papers in the present volume are arranged in such a way that they correspond with the aforementioned topics. (Thus the order implies no evaluation of the various papers!). This subdivision should not be misunderstood, and we would like to call all the papers to the attention of the reader - whatever his specialization. It seems to be a feature of our time that scientists build or reconstruct bridges from their own fields to other ones. We hope that the workshop as well as the proceedings add a brick to these bridges.

We wish to thank Dr. Höppner from Akademie-Verlag for his help and cooperation.

Jena/Glasgow, December 1988

The editors

T A B L E O F C O N T E N T S

Digital Geometry

Computational Geometry

Integral Geometry and Stochastic Geometry

Stereology

Geometry and other contributions to Image Processing

Last Minute Papers

Discrete Straight Line Segments: Parameters, Primitives and Properties

Leo Dorst [1]

Robotics and Flexible Automation, Philips Laboratories, Briarcliff Manor, NY 10510, USA

Abstract

This paper is a terse overview of the main results of my research in discretized straight lines. It presents a formalism in which estimators for properties of straight lines, on the basis of their discretization, can be formulated and assessed. For the important proper 'length' the results are given in greater detail, and recommendations for the appropriate length estimators for different circumstances are given.

1 Introduction

In digital image analysis, the aim is to measure some aspects of the continuous world on the basis of digital images. In many applications, geometrical properties are foremost among the aspects to be measured. Therefore it is of interest to study how accurately these continuous parameters can still be determined from the digital data. The issues are non-trivial; this paper presents some of the first results, for discretized straight lines.

A typical problem is the following. Assume we have a 2-dimensional disk in the world, black against a white background, described by a function $f(x,y)$ which is 0 inside the object and 1 outside. The goal is to measure the circumference of the disk. By means of a camera and a digitizer, this object is represented in a discrete array as a function $g(i,j)$, which may assume values in the continuous*range* $[0,1]$ (when properly scaled) due to the imperfections of optics and digitization. From this function, we have to derive a good estimate of the circumference - exact measurement has become impossible.

The measurement could be performed by first estimating the original object $f(x,y)$ on the basis of $g(i,j)$, and measuring the circumference of that estimate $\hat{f}(x,y)$. Alternatively, it could be performed by finding a good estimate of the circumference based on the discrete data directly. For instance, one could threshold $g(i,j)$ at value $\frac{1}{2}$, approximate the contour by a chaincode, count the number of odd chaincode elements n_o and even chaincode elements n_e, and estimate the circumference by estimator $L = n_e + n_o\sqrt{2}$.

The first method of measurement requires the estimating of a continuous function $f(x,y)$ on the basis of a deformed and sampled version $g(x,y)$. This is amenable to classical image reconstruction techniques. However, the circumference of the best reconstruction is not necessarily the best reconstructed circumference. The second estimation technique is more commonly used in image processing, mainly because it is simpler and faster.

Inspecting formulas used to determine the length of a discrete straight line, we found that many are based on discrete geometry alone, without evaluating their performance in actually coming up with a good value for the length of the original, continuous line. In fact, when we started this study, optimal estimators were not known, and the 'estimator' mentioned above, $L = n_e + n_o\sqrt{2}$, was commonly used.

In this paper we study the problem of measuring properties such as length, slope, angle, etc. of a straight fragment of a contour, discretized ideally (no distortion, no noise). This is admittedly a long way off from the general geometric parameter estimation problem outlined before. But we believe that this problem, and our way of solving it, sheds some light on the difficulties involved in the general problem.

The paper is organized as follows. First, in Section 2 and 3, discretized line segments are described, and several characterizations of them by means of *parameters* are developed. Section 4 describes the inverse process, of finding the *primitives*, given the parameters. Then, in Section 5, estimators are defined for *properties* of the continous line segment, taking these characterizations as input. Various criteria and digitizations are studied. In Section 6 estimators are developed for the important property 'length', and in Section 7 the results of the different estimators are compared, showing that the best-known estimators are not always the best.

2 Digitization

Suppose we have a continuous straight boundary with equation

$$y = \alpha x + e, \tag{1}$$

with the object lying in the direction of negative y. Without loss of generality, we can assume:

$$0 \leq \alpha \leq 1, \quad 0 \leq e < 1 \tag{2}$$

[1]Work done when with Pattern Recognition Group, Applied Physics, Delft University of Technology, Delft, The Netherlands

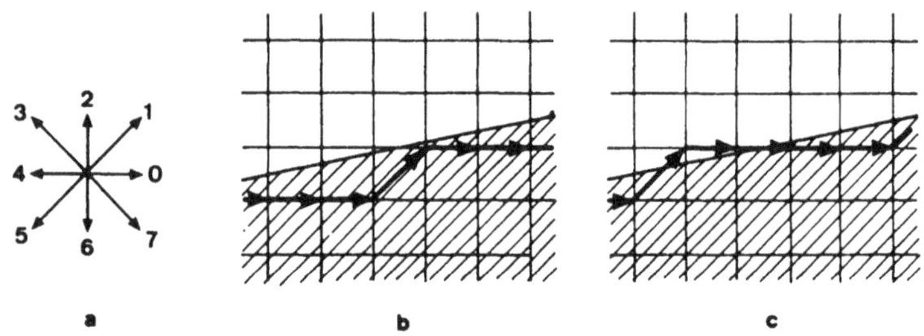

Figure 1: a: Chaincodes, b: OBQ digitization, c: GIQ digitization.

Digitizing the line of eq.(1) by 'object boundary quantization' (OBQ) [8], we obtain the discrete points:

$$(i, y_i) = (i, \lfloor \alpha i + e \rfloor) \tag{3}$$

Here $\lfloor . \rfloor$ indicates the floor function, or 'rounding down'. Another common method of digitization, used for lines rather than for boundaries, is 'grid intersection quantization' (GIQ) [8], which gives the points:

$$(i, y_i') = (i, [\alpha i + e]) \tag{4}$$

where $[.]$ indicates the rounding-off function. Since we have that $[x] = \lfloor x + \frac{1}{2} \rfloor$, we need only study OBQ, noting that results apply to GIQ when we rewrite $e \leftarrow e + \frac{1}{2}$.

Discrete straight lines are commonly represented by their chaincode string C, of which the i-th element, c_i is:

$$c_i \equiv y_i - y_{i-1} \tag{5}$$

This is the representation we use. Straight strings of lines with the limitations of eq.(2) consist only of code elements 0 and 1. (see Fig.1).

For the remainder of this paper, we consider *finite* strings, consisting of n elements, and we consider such a string as a finite part of an infinitely long string. We do not consider a finite string as the digitization of a continous line segment - that would require special treatment of the end points.

The chaincode representation of straight lines has interesting properties, and it can be determined whether a given chaincode string could have been the discretization of a straight line, by checking the so-called 'linearity conditions' (which bear interesting relations to number-theoretical issues in the approximation of real numbers by rational numbers), see [14,3].

3 Parameters

The chaincode string representation is not yet convenient to use for estimation: formulas for estimators do not take strings as input, but numbers characterizing the string. As an example, the length estimator given in the introduction was based on the number of odd and even chaincodes. Four characterizations are studied.

3.1 (n)-characterization

In this characterization, the string C is characterized by the number of chaincode elements n. All strings with the same number of elements are thus characterized by the same 1-tuple n. Among these, there are many non-straight strings.

3.2 (n_e, n_o)-characterization

The string C is characterized by the total number of even code elements n_e and the number of odd code elements n_o. The endpoint of the discrete line segment corresponding to a straight string C starting at (0,0) is $(n_e + n_o, n_o)$. Thus all discrete line segments with the same endpoints have the same tuple (n_e, n_o). Moreover, there are non-straight strings between the endpoints that also have the same tuple (namely all strings with only code elements 0 and 1).

3.3 (n_e, n_o, n_c)-characterization

The parameters n_e and n_o have the same meaning as before. The parameter n_c is the 'corner count', the number of transitions between odd and even elements, first introduced in [13]. It is the number of occurrences in the string where $c_i \neq c_{i-1}$. It can be shown that strings have the same tuple (n_e, n_o, n_c) if and only if they have the same discrete points at the columns $i = 0, 1, (n-1), n$ (see [13]).

3.4 (n, q, p, s)-characterization

This is a tuple of parameters first introduced in [2]. It is defined by:

$$\begin{cases} n \textit{ is the number of elements of } C \\ q = \min_k\{k \in \{1,2,...,n\} \mid k = n \vee \forall i \in \{1,2,...,n-k\} : c_{i+k} = c_i\} \\ p = \sum_{i=1}^{q} c_i \\ s : s \in \{0,1,...,q-1\} \wedge \{1,2,...,q\} : c_i = \lfloor \frac{p}{q}(i-s) \rfloor - \lfloor \frac{p}{q}(i-s-1) \rfloor \end{cases} \tag{6}$$

These rather complex definitions lead to a tuple of parameters that bears a one-to-one relation to the chaincode string C. The i-th element c_i of C is given by:

$$c_i = \lfloor \frac{p}{q}(i-s) \rfloor - \lfloor \frac{p}{q}(i-s-1) \rfloor. \tag{7}$$

This equation shows how the parameters p, q and s should be interpreted: $\frac{p}{q}$ is a rational approximation to α (in fact it is the simplest approximation, in terms of having minimum denominator), s is a 'phase shift', related to e. For a proof and more details, see [2,4,10].

By eq.(7), strings with the same tuple (n, q, p, s) are necessarily the same string. Such characterization with a one-to-one correspondence between tuple and string is called a *faithful* characterization. Also by eq.(7), strings for which a tuple (n, q, p, s) exists are necessarily straight.

In [10], an interesting expression for s is given, and the number of straight strings of n elements is derived on the basis of the (n, q, p, s)-characterization, showing that a faithful characterization can have many uses.

4 Primitives

Let us denote a continuous line by ℓ, its digitization (in the form of a chaincode string) by $C = \mathcal{C}\ell$, the characterization of C by $T = \mathcal{T}C = \mathcal{T}\mathcal{C}\ell$. Denote the set of strings with a given T by $\mathcal{T}^{-1}T$, and the set of lines with a given C by $\mathcal{C}^{-1}C$. The set $\mathcal{C}^{-1}\mathcal{T}^{-1}T$ for a given tuple T is thus by definition the set of continuous straight lines ℓ whose tuple $\mathcal{T}\mathcal{C}\ell$ equals T. These sets are conveniently expressed in the parametrization $\ell = (e, \alpha)$ for a given n.

Expressions for these sets (also called 'regions') are given in [4]. In (e, α)-space, they look like collections of rhomboids. For the (n)-characterization and the (n_e, n_o)-characterization, these can be found easily. The expressions for $\mathcal{C}^{-1}\mathcal{T}^{-1}(n_e, n_o, n_c)$ were first given in [13]. The expressions for $\mathcal{C}^{-1}\mathcal{T}^{-1}(n, q, p, s)$ were first given in [2], and are derived more directly in [12], where they are called 'facets'. Fig.2 plots the regions for all strings with $n = 6$ consisting of codes 0 and 1, in the four characterizations introduced in Section 3.

It is seen from this figure that the finer the characterization is, the finer is the tesselation of (e, α)-space: more lines have become distinguishable. Since the (n, q, p, s)-characterization is a faithful characterization, the regions of this characterization are also the equivalence classes of lines still distinguishable after digitization. Each string has its own 'domain'.

These domains are so fundamental in the study of discrete straight lines that it is worth repeating the expressions derived in [2]. In object boundary quantization, $\mathcal{C}^{-1}\mathcal{T}^{-1}(n, q, p, s)$ is given by:

$$\begin{array}{ll} 1) & \frac{p_-}{q_-} < \alpha < \frac{p_+}{q_+} \\ 2) & \lceil \frac{sp}{q} \rceil - s\alpha \leq e < \lceil \frac{L(t)p+1}{q} \rceil - L(t)\alpha \quad \text{if } \frac{p}{q} \leq \alpha < \frac{p_+}{q_+} \\ & \lceil \frac{L(s)p}{q} \rceil - L(s)\alpha \leq e < \lceil \frac{tp+1}{q} \rceil - t\alpha \quad \text{if } \frac{p_-}{q_-} \leq \alpha < \frac{p}{q} \end{array} \tag{8}$$

where $t : tp = (sp-1) \bmod q$, $L(x) = x + \lfloor \frac{n-x}{q} \rfloor$, $p_+ = \lceil \frac{pL(t)+1}{q} \rceil - \lceil \frac{sp}{q} \rceil$, $p_- = \lceil \frac{pL(s)}{q} \rceil - \lceil \frac{tp+1}{q} \rceil$, $q_+ = L(t) - s$ and $q_- = L(s) - t$. A proof can be found in [2], an improved proof in [4].

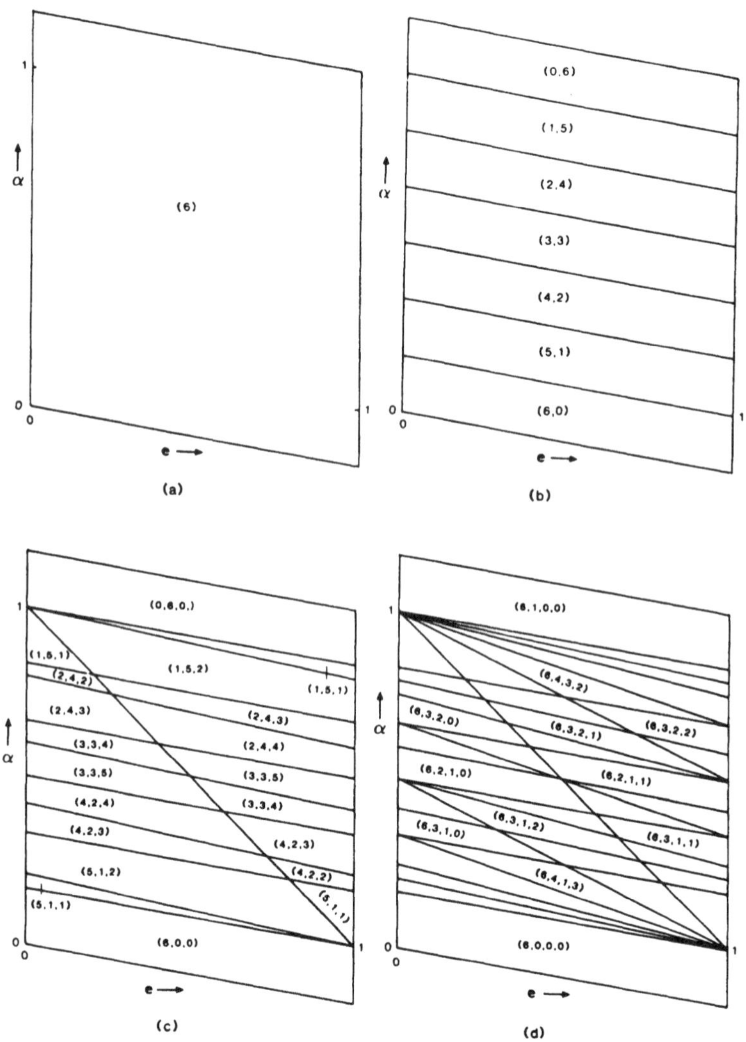

Figure 2: Regions in (e, α) of all strings with $n = 6$, for the 4 characterizations.

5 Properties

Denote the property of ℓ to be estimated by $f(\ell)$, and the estimator by $g(T)$. All lines ℓ in $\mathcal{C}^{-1}\mathcal{T}^{-1}T$ have the same value for T, and hence the property $f(\ell)$ for all these lines is estimated by the same value $g(T)$.

The art of designing good estimators, given $\mathcal{C}$ and $\mathcal{T}$, is to minimize a measure of error between $f(\ell)$ and $g(T)$, for all $\ell \in \mathcal{C}^{-1}\mathcal{T}^{-1}T$. Two types of estimator are considered here: Best Linear Unbiased Estimators (BLUEstimators) and Most Probably Original (MPO) Estimators.

A *BLUE estimator* is a linear estimator (i.e. it is an average over values of the function $f(\ell)$ over some range of ℓ), that minimizes the mean square error (MSE) between $f(\ell)$ and $g(\mathcal{T}\mathcal{C}\ell)$ over all ℓ considered. It is given by:

$$g_{BLUE}(T) = \int_{\mathcal{C}^{-1}\mathcal{T}^{-1}T} f(\ell)p(\ell)d\ell \tag{9}$$

where $p(\ell)$ is the probability density of lines. In words, the BLUEstimator for f given T is the average of f over the region corresponding to T.

The *MPO estimator* for a property f is the most probable original value for the property, given T. In formula:

$$g_{MPO}(T) = f(\text{argmax}\{p(\ell)|\ell \in \mathcal{C}^{-1}\mathcal{T}^{-1}T\}) \tag{10}$$

where $\text{argmax}\{p(\ell)\}$ indicates the value of ℓ maximizing $p(\ell)$ in the given range.

In [5] such estimators are developed for arbitrary properties of straight line segments. In [4] different types of estimators are treated. Here we focus on length estimators as an important example.

6 Length Estimators

The property 'length' of the continuous straight line segments considered (those spanning n columns of the grid) is given by:

$$f(e, \alpha, n) = n\sqrt{1+\alpha^2}. \tag{11}$$

Note that this is independent of e.

We will develop estimators for this property. As a measure for the accuracy of a length estimator L we use $RDEV(L, n)$, defined as the square root of the mean square error of L, averaged over all straight strings of n elements 0 and/or 1, divided by n:

$$RDEV(L(T), n) = \frac{1}{n}\sqrt{\sum_T \int_{\mathcal{C}^{-1}\mathcal{T}^{-1}T} (L(T) - f(\ell))^2 p(\ell)d\ell}. \tag{12}$$

The normalization by n permits interpretation of $RDEV(L, n)$ as the relative error in the length measurement of all line segments with a projected length of unity, when the sampling density is n^2 per square unit.

Many length estimators can be developed (see [4]). Here we will consider the MPO- and BLUE- estimators, ordered by characterization. Then, we introduce and discuss a different type of estimator, important for fast measurement: the *simple* length estimators.

6.1 (n)-characterization

For this characterization, the MPO length estimator is given by:

$$g_{MPO}(n) = n \tag{13}$$

The BLUEstimator is:

$$g_{BLUE}(n) = \frac{\pi}{4}\sqrt{2}n = 1.111n. \tag{14}$$

(see [6]). For both estimators, the error $RDEV$ tends to a constant as $n \to \infty$. This implies that the estimation does not get more accurate than some accuracy, even not when the sampling density becomes infinite. For $g_{MPO}(n)$, the asymptotic error is 16%, for $g_{BLUE}(n)$ it is 11%.

6.2 (n_e, n_o)-characterization

For the (n_e, n_o)-characterization, the MPO length estimator is given by:

$$g_{MPO}(n_e, n_o) = \begin{cases} \frac{1}{2}\sqrt{3} & \text{if}(n_e, n_o) = (0,1) \\ \sqrt{(n_e+n_o)^2 + n_o^2} & \text{elsewhere} \end{cases} \tag{15}$$

Notice that this is in fact the Euclidean distance between the endpoints of the discrete line segment (except for the case $(n_e, n_o) = (0,1)$). This seems a reasonable enough estimator of the length of the discrete line segment. Surprisingly, it is *not* the most accurate length estimator! It can be shown that the asymptotic behavior of the accuracy as measured by $RDEV$ is $O(n^{-1})$ (see [4]).

The BLUEstimator for the (n_e, n_o)-characterization is more complicated. Let us put, for convenience, $n = n_e + n_o$ and $m = n_o$. Then:

$$\begin{aligned} g_{BLUE}(n_e, n_o) &= \frac{\sqrt{2}}{n}\{\frac{m+1}{n}\text{atan}(\frac{m+1}{n}) - 2\frac{m}{n}\text{atan}(\frac{m}{n}) + \frac{m-1}{n}\text{atan}(\frac{m-1}{n}) \\ &- \frac{1}{2}\log(1+(\frac{m+1}{n})^2) + \log(1+(\frac{m}{n})^2) - \frac{1}{2}\log(1+(\frac{m-1}{n})^2)\} \end{aligned} \quad (16)$$

The asymptotic $RDEV$ of this estimator is also $O(n^{-1})$. A better order than this is not to be expected from any estimator based on the (n_e, n_o)-characterization .

6.3 (n_e, n_o, n_c)-characterization

The MPO length estimator for the (n_e, n_o, n_c)-characterization have not been computed, due to the complex form of the regions. The asymptotic order can be computed, though, and is found to be $O(n^{-1})$ (see [4]).

The BLUEstimator was first computed in [13]. The resulting formulas are rather complex and are not repeated here. The asymptotic accuracy is $O(n^{-1})$.

6.4 (n, q, p, s)-characterization

The MPO-estimator for the (n, q, p, s)-characterization is simply (see [4]):

$$g_{MPO}(n, q, p, s) = n\sqrt{1+(\frac{p}{q})^2}. \quad (17)$$

This is independent of s (which is because the original f in eq.(1) is independent of e).

Though similar in form to $g_{MPO}(n_e, n_o)$, this estimator is more accurate: the asymptotic error is $O(n^{-\frac{3}{2}})$. The reason is that $\frac{p}{q}$ is a better estimate of the slope α of the real line than $\frac{n_o}{n_e+n_o}$. The latter is an approximation by a fraction with fixed denominator $n = n_e + n_o$, of which there are only a total number of n. The former approximates α by an irreducible fraction with denominator between 1 and n, of which there are $O(n^2)$ (see [9]).[2]

The BLUEstimator for the (n, q, p, s)-characterization was given in [5]. As with the BLUEstimator for the (n_e, n_o, n_c)-characterization , the expressions are involved (in fact very similar!) and are not repeated here. The BLUEstimator for the (n, q, p, s)-characterization is the most accurate estimator for the length of a discrete line that can be constructed (in terms of minimizing the mean square error). The asymptotic order is $O(n^{-\frac{3}{2}})$, and by calculation we have established the coefficient of proportionality to be 0.34.

6.5 Simple Length Estimators

We will call estimators that are linear in the parameters of the characterization *simple* estimators, indicated by $L_S(T)$. The optimal coefficients in the linear function are to be determined by an optimization criterion, for which we use minimization of the mean square error over all $\ell \in C^{-1}T^{-1}(T)$. The simple estimators considered are given in Fig.3. Historically, L_F was first given in [7], L_G in [8], L_K in [11] and L_C in [13]. More on the historical development can be found in [6].

Note that $L_0(n) = g_{MPO}(n)$ and $L_1(n) = g_{BLUE}(n)$. Of the estimators depicted in the table, $L_0(n)$, $L_F(n_e, n_o)$ and $L_N(n_e, n_o, n_c)$ are straight-forward estimators, counting 'grid moves' (code element 0) as length 1, 'diagonal moves' (code element 1) as length $\sqrt{2}$ and 'knight's moves' (neighboring codes 0 and 1) as $\sqrt{5}$. They are biased, and therefore have a large $RDEV$. The estimators $L_1(n)$, $L_K(n_e, n_o)$ and $L_C(n_e, n_o, n_c)$ are the optimal linear estimators, linear estimators for which the coefficients are chosen to minimize the MSE. $L_G(n_e, n_o)$ minimizes the MSE for strings with $n = 1$, or for longer strings with uncorrelated elements.

Fig.4 plots $RDEV$ for these estimators as a function of n. Note that these simple estimators have a *limited accuracy*: even with a sampling density that goes to infinity, they do not become more accurate, but the error converges to an asymptotic constant value. Therefore, if one wants to measure the length of a contour with an accuracy of 1% by a simple estimator, this is only possible with L_C, and then only if one samples with a sampling density of about 40 points along the length to be measured. Note that Fig.4 gives average values for the error (see eq.(12)) – some directions have a worse error, specifically those along the grid axes.[3]

[2] The distribution of these fractions is not uniform, and that is the reason that one does reach an accuracy better than $O(n^{-\frac{3}{2}})$.

[3] It is interesting to note that the direction $\alpha = \text{atan}(\sqrt{2}-1)$ is the most accurate: it pays to tilt the camera to that angle relative to the contour to be measured!

estimator		$RDEV(g(T),\infty)$
$L_0(n)$	$= 1.000n$	16 %
$L_1(n)$	$= 1.111n$	11 %
$L_G(n_e, n_o)$	$= 1.059n_e + 1.183n_o$	8.0 %
$L_F(n_e, n_o)$	$= 1.000n_e + 1.414n_o$	6.6 %
$L_K(n_e, n_o)$	$= 0.945n_e + 1.346n_o$	2.6 %
$L_N(n_e, n_o, n_c)$	$= 1.000n_e + 1.414n_o - 0.089n_c$	1.1 %
$L_C(n_e, n_o, n_c)$	$= 0.980n_e + 1.406n_o - 0.091n_c$	0.8 %

Figure 3: The simple estimators.

7 Recommendations for Length Estimators

Fig.4 shows the accuracy all length estimators treated as a function of n. The best estimator is $g_{BLUE}(n, q, p, s)$, as expected. Its asymptotic accuracy is $0.34n^{-\frac{3}{2}}$. The fact that this is the best estimator can be interpreted in terms of the required sampling density to reach a certain accuracy. This leads to the following:

Trade-off Theorem
For length measurement of discrete straight lines, the sampling density d (per unit length) is related to the asymptotically best achievable estimation error percentage p by:

$$d > 10p^{-\frac{2}{3}} \tag{18}$$

Equality is reached with the length estimator $g_{BLUE}(n, q, p, s)$ only.

Thus for an error of 1%, one needs $d > 10$, when using $g_{BLUE}(n, q, p, s)$, and for an error of 0.1%, one needs $d > 50$.

On the basis of Fig.4, and considerations of computational efficiency, we can give some recommendations as to which estimator to use. We come to the following recommended estimators:

- $L_1(n)$ can be used in extremely time-critical situations, or in simple image analysis (count the contour pixels and multiply by 1.111). Accuracy up to 11%. Use instead of $L_0(n)$.
- $L_K(n_e, n_o)$ can be used in time-critical situations or when no high accuracy is required. Accuracy up to 2.6%. To be used instead of $L_G(n_e, n_o)$ and $L_F(n_e, n_o)$ (the latter is the familiar '$n_e + \sqrt{2}n_o$').
- $L_C(n_e, n_o, n_c)$ is the 'corner count' estimator. For normal use: simple to compute, reasonable accuracy, up to 0.8%. Comparable to optimal result for short strings $n < 10$.
- $g_{MPO}(n_e, n_o)$ is the Euclidean distance between the endpoints of the discrete line segment. It is easy to compute. The accuracy is $0.17n^{-1}$.
- $g_{MPO}(n, q, p, s)$ for use when high accuracy is desired – it is only slightly less accurate than the much more involved optimal estimator $g_{BLUE}(n, q, p, s)$. Asymptotic accuracy $0.46n^{-\frac{3}{2}}$.

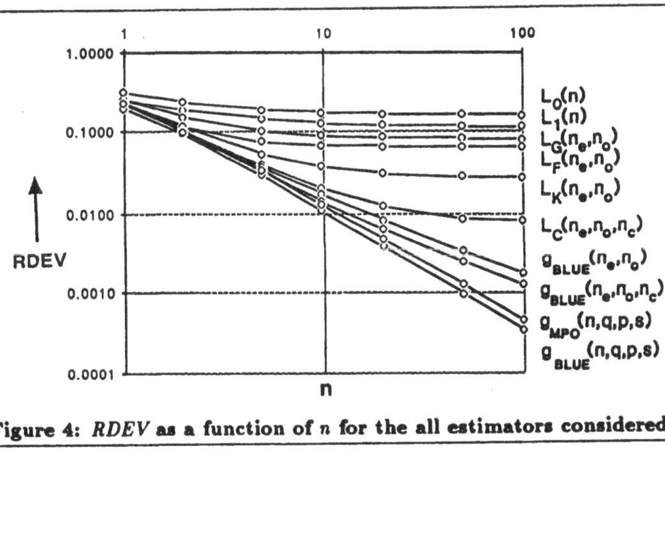

Figure 4: *RDEV* as a function of n for the all estimators considered.

It should be emphasized that these results apply only to strings that are the digitization of straight lines (so-called 'straight strings'). We have run some experiments on the performance of the estimators on non-straight strings. The results can be found in [6].

8 Conclusion

This research in discrete straight lines was started to derive accurate estimates for their properties. This goal was achieved for length estimators with the computation and evaluation given above. For other properties than length, computations can be found in [5,4], but no detailed comparisons have been made.

In order to perform this analysis, many fundamental issues had to be (un)covered.

First, we had to find a faithful characterization of a straight string, since only if we use all information present in the string can we expect to find optimal estimators. That characterization is (n, q, p, s). It was first given in [2]. The representation is being used by others now, for instance in [10] and [1].

Then, we needed to know which continuous lines could be digitized to a given string. That set of lines, the 'domain' of the string, was first given in [2], with an awkward proof. A more insightful proof was given in [12], which formed the basis of a new proof in [4]. In [12], the domains (called 'facets') are used to rapidly find the chaincode string corresponding to a given line.

From the domains, optimal estimators could then be derived in a straight-forward manner. BLUEstimators for arbitrary properties (not just line length) are given in [5]. Other optimal estimators, not just BLUE, for arbitrary properties are given in [4].

For the comparison of different length estimators, it was useful to develop the notion of a characterization. Equivalence classes of lines for each characterization considered here are given in [4]. The asymptotic accuracy of the optimal estimators for each characterization shows the influence of the variables used: for the (n)-characterization it is $O(1)$, for the (n_e, n_o)- and (n_e, n_o, n_c)-characterization it is $O(n^{-1})$, and for the (n, q, p, s)-characterization it is $O(n^{-\frac{3}{2}})$, which is the optimal accuracy. The use of the (n_e, n_o, n_c)-characterization lies in the very acceptable accuracy of the simple estimator $L_C(n_e, n_o, n_c) = 0.980n_e + 1.406n_o - 0.091n_c$.

The results of the present paper are found in more detail in [6]. That paper also consider length estimators for non-straight strings. In [5,4], estimators for properties other than length are given.

References

[1] C.A. Berenstein, L.N. Kanal, D. Lavine, *A Geometric Approach to Subpixel Registration Accuracy*, Computer Vision, Graphics and Image Processing, Vol.40, No.3, pp.334-360, 1987.

[2] L. Dorst and A.W.M. Smeulders, *Discrete Representation of Straight Lines*, IEEE Transactions on Pattern Analysis and Machine Intelligence, Vol. PAMI-6, No.4, pp.632-639, 1984.

[3] L. Dorst and R.P.W. Duin, *Spirograph Theory, a Framework for Calculations on Digitized Straight Lines*, IEEE Transactions on Pattern Analysis and Machine Intelligence, Vol. PAMI-6, No.5, pp.632-639, 1984.

[4] L. Dorst, *Discrete Straight Line Segments: Parameters, Primitived and Properties*, Ph.D. Thesis, Technological University Delft, 1986.

[5] L. Dorst and A.W.M. Smeulders, *Best Linear Unbiased Estimators for Properties of Digitized Straight Lines*, IEEE Transactions on Pattern Analysis and Machine Intelligence, Vol. PAMI-8, No.2, pp.276-282, 1986.

[6] L.Dorst and A.W.M. Smeulders, *Length Estimators for Digitized Contours*, Computer Vision, Graphics and Image Processing, Vol.40, pp.311-333, 1987.

[7] H. Freeman, *Boundary Encoding and Processing*, In: Picture Processing and Psychopictorics (B.S. Lipkin and A. Rosenfeld, eds.), Academic Press, New York, pp.241-266, 1970.

[8] F.C.A. Groen and P.W. Verbeek, *Freeman Code Propbabilities of Object Boundary Quantized Contours*, Computer Graphics and Image Processing, Vol.7, pp.391-402, 1978.

[9] G.H. Hardy and E.M. Wright, *An introduction to the theory of numbers*, Oxford, 5th edition, 1979.

[10] J. Koplowitz, *A Robust Filtering Algorithm for Subpixel Accuracy Reconstruction of Chain Coded Line Drawings*, IEEE Transactions Pattern Analysis and Machine Intelligence, Vol.9, No.3, pp.451-457, 1987.

[11] Z. Kulpa, *Area and Perimeter Measurement of Blobs in Discrete Binary Pictures*, Computer Vision and Image Proessing, Vol.6, pp.4334 454, 1977.

[12] M.D. McIlroy, *A Note on Discrete Representation of Lines*, AT&T Technical Journal, Vol.64, No.2, 1985.

[13] A.M. Vossepoel and A.W.M. Smeulders, *Vector Code Probability and Metrication Error in the Representation of Straight Lines of Finite Length*, Computer Graphics and Image Processing, Vol.20, pp.347-364, 1982.

[14] L.-D. Wu, *On the Chain Code of a Line*, IEEE Transactions on Pattern Analysis and Machine Intelligence, Vol.PAMI-4, No.3, pp.347-353, 1982.

THE STRUCTURE OF BOUNDARIES OF DIGITAL SETS WITH APPLICATION TO PARALLEL THINNING

Ulrich Eckhardt
Institut für Angewandte Mathematik
Universität Hamburg
Bundesstraße 55
D-2000 Hamburg 13

Abstract. When points are removed in parallel from a digital set, the topology of the latter might change. In this paper, those points which can be safely eliminated in parallel are characterized by means of the local structure of the boundary of the set. This characterization will be helpful for understanding the properties of parallel thinning algorithms and also of thinning algorithms working on boundary data.

1. Introduction

There exists a large number of publications on thinning of binary images. In spite of this fact, however, there remain many problems and open questions in this area. Specifically, thinning methods are not well-defined, i.e. the result of their application depends on the implementation. Moreover, these methods are usually not invariant with respect to rotations, translations and reflections of the digital plane. Most proposals for parallel implementation of thinning algorithms are ad-hoc in nature and thus there seems a need for closer theoretical investigation.

The aim of this paper is to show that there is a very close relationship between the local structure of the boundary of a digital set and the property of points on the boundary to be suited for parallel elimination from the set without affecting the topology of the latter. Indeed, attempts to formulate a boundary-oriented thinning method nearly automatically leads to the investigation of the structure of the boundary with similar results as presented here |2|.

2. Basic Definitions

The *digital plane* is the set of all points in the plane having integer coordinates. A *digital set* S is any finite subset of the digital plane. For convenience, points in S will be termed *black* or *foreground points*, while those of the complement will be termed *white* or *background points*.
Each point P in the digital plane has eight *neighbors*, which are numbered 0 to 7 according to the following scheme

3	2	1
4	P	0
5	6	7

The notation $N_i(P)$ refers to the neighbor of P having number i. N(P) is the set of all neighbors of P (without P itself) and is termed the *neighborhood* of P. Neighbors with even numbers are *direct neighbors* of P or *4-neighbors*; neighbors having odd numbers are *indirect neighbors* of P. All neighbors together are sometimes called *8-neighbors*.

By means of the neighbor-relation we are able to define a topology in the digital plane |5|. As usual we assume the foreground to be equipped with the 8-topology, i.e. two points in S are connected if there exists a sequence of points in S starting at one of the given points and ending at the other such that each two successive points in the sequence are 8-neighbors of each other. The background is analogously equipped with the 4-topology. From this definition it becomes clear what is meant by foreground and background *connection components*.

For any point P the *connection number* C(P) is the number of black 8-components in N(P). A *simple point* is a point of S having connection number one and at least one white direct neighbor. An *end point* is a point in S having exactly one black neighbor.

Thinning of a set consists in removing from it simple points (but usually not end points). Rosenfeld |5| proved that the number of connection components of the set and also of the complement remains unchanged under elimination of simple points. Thus, thinning is suitable for simplifying a set without changing its topological structure. If no more simple points remain in a set, it is said to be a *skeleton*.

3. The Boundary of a Digital Set

A *boundary point* of a digital set is a black point having at least one direct white neighbor. Points of a set which are not boundary points are called *interior points*. The set of all interior points is the *kernel* of a set.

It is well-known (see e.g. |4|) that each pair (P,Q) consisting of a black point P and a white point Q which is a 4-neighbor of P determines uniquely a *component of the boundary* which may be found by applying the *boundary following algorithm* (see |4| or |5|) to (P,Q). In this algorithm for each pair (P,Q) as above the *successor pair* is defined as follows:

Let R be the point in N(P) which follows Q in counterclockwise direction. If R is black, then (R,Q) is the successor pair of (P,Q).
If R is white, let S be the point in N(P) following R in counterclockwise direction. If S is black, (S,R) is the successor pair, otherwise (P,S) is the successor pair of (P,Q).

By the successor-relation, the boundary components become *oriented*. The *predecessor pair* of a boundary pair can be found by a similar rule and thus it is uniquely determined too.

A boundary point having an 8-neighbor which is an interior point is said to be a point on the *inner boundary*. In a similar way as above we can construct a successor pair on the inner boundary for any pair (P,Q) consisting of a point P on the boundary and a point Q in the kernel such that P and Q are 8-neighbors of each other:

Let R be the point in N(Q) which follows P in N(Q) and is black. If R is no interior point, then (R,Q) is the successor pair, otherwise (P,R) is the successor pair of (P,Q).

Also here, the construction process is well-defined and yields a connection component of the inner boundary surrounding a kernel 8-component.

Generally, if it is clear which white partner is meant, we simply speak of a successor or a predecessor of a point on the boundary and similarly for the inner boundary.

Yokoi, Toriwaki and Fukumura |7| observed that the connection number of a point on the boundary coincides with the number of times this point occurs on different pieces of the boundary. Pavlidis |4| correlated certain properties of simple points with deviations of the boundary at these points from being an 8-curve. Here it is shown that any point with the property that the boundary and the inner boundary coincide in it (as oriented sets) is an ideal candidate for parallel elimination.

4. Perfect Points

A point P on the inner boundary of a digital set is said to be *perfect* if there exists an interior point in N(P) such that all direct neighbors of P which are not neighbors of this interior point, are white. This definition differs slightly from the definition in |1|. Points on the boundary which are simultaneously simple and perfect can be eliminated from the set independently from each other and thus simultaneously whithout changing the topology of the set |1|. On the other hand, these points can be characterized by their position with respect to both (oriented) boundaries:

Theorem 1: Given a point P on the inner boundary. If the successor of P on the boundary coincides with the successor of P on the inner boundary, and if the same holds for the predecessors, then P is simple and perfect.

The proof of this theorem is performed by discussing for a point P all possibilities for the successors and predecessors on both boundaries |1|.

Although the converse of the theorem is not generally true, we may state:

Theorem 2: Given a simple perfect point P. If the successor of P on the boundary is perfect then it is also the successor of P on the inner boundary. An analogous assertion holds for the predecessors.

This theorem can be proved by investigation of all possible neighborhood configurations for simple and perfect points |1|.

5. Connection Numbers

When a point is deleted from a set, the connection numbers of its neighbors may change. Define the *total connection number* $C(S)$ of a digital set S to be the sum of all connection numbers of points in this set and the *average connection number* $\mu(S)$ to be the total connection number of S divided by the number of points contained in S. One might expect the total connection number of a set to decrease and the average connection number to increase under thinning. Hence, it is of interest to investigate the change of these numbers when a simple point is removed from a set.

Theorem 3: Given a point P with connection number $C(P)$. If a simple black neighbor Q of P is eliminated then $C(P)$ will become smaller if and only if Q is an end point. One has in this case

$$C(S - \{P\}) = C(S) - 2$$

$$\mu(S - \{P\}) \leq \mu(S) \quad \text{if and only if } \mu(S) \leq 2.$$

If Q is an indirect neighbor of P then $C(P)$ is not changed if and only if Q is no end point.

The proof of this theorem is straightforward |1|. The assertion of the theorem is the theoretical basis for excluding end points from elimination.

When a single point is eliminated, we get:

Theorem 4: Let P be a simple point of a digital set S, but no end point. Then always

$$\mu(S) \leq \mu(S - \{P\}).$$

Furthermore,

1. If P is on the inner boundary, then
   - 1.1. The successor of P on the inner boundary is equal to the successor of P on the boundary and the same holds for the predecessors. Then P is perfect and

     $$C(S - \{P\}) = C(S) - 1.$$

   - 1.2. The successors or the predecessors of P are different.
     - 1.2.1. If P is perfect then there exists a direct neighbor Q of P such that Q is simple but not perfect with respect to S but it is not simple with respect to $S - \{P\}$. Moreover,
       $$C(S) \leq C(S - \{P\}).$$
     - 1.2.2. If P is not perfect then there exists a direct neighbor Q of P such that $C(Q)$ is increased by elimination of P. Furthermore,
       $$C(S) \leq C(S - \{P\}).$$

2. If P does not belong to the inner boundary and is no end point then
   - 2.1. P has three direct black neighbors or it has one direct black neighbor and three black neighbors. Then there exists a direct neighbor Q of P such that $C(Q)$ is increased by elimination of P

and

$$C(S) \leq C(S - \{P\}).$$

2.2. P has two direct black neighbors or it has one direct black neighbor and two black neighbors. Then

$$C(S - \{P\}) = C(S) - 1.$$

For a proof of this theorem, see |1|.

Any point on the boundary whose predecessors and successors coincide (case 1.1 of theorem 4) can be removed without any precautions independently of other simple points. If the successors or the predecessors are not equal, but the point is perfect (case 1.2.1), a *collision* occurs under parallel elimination, i. e. if P is eliminated, the originally simple point Q will become a skeleton point. Hence P and Q cannot be eliminated simultaneously. The only invariant decision in this case is to eliminate the perfect partner. In all other cases (except case 2.2) a collision is possible. From knowledge of the neighborhood of a point alone it is not possible to decide whether one of its direct neighbors is simple or not. Even if this information is available, e.g. by a suitable label, then it is not possible to see whether a direct neighbor will remain simple after elimination of the point under consideration.

It is interesting to note that case 2.2 covers the counterexample of Lü and Wang |3| showing that Rutovitz' method for parallel thinning is not universally valid and also the counterexample of Tamura |6|:

```
         x x
       x x
     x x
   x x
  .
 .
```

A spike of this form will be deleted by any known method for parallel thinning if no special precautions are provided. Note that all points of this set are simple but none is perfect.

When all simple and perfect points of a set are eliminated, the total connection number of this set is not necessarily decreased. Consider the following example:

```
    x x x                      x x
x x x x          ——>     x x x
    x x x                    x x x
    x                        x
```

Here C(S) = 12. After elimination of all simple and perfect points we get a set with total connection number 13.

6. Conclusions

Parallel elimination of simple and perfect points from a digital set is advantageous from a theoretical point of view. Specifically it leads to a well-defined method for parallel thinning which exhibits all motion invariances of the digital plane.

Here it was shown that simple and perfect points can be characterized by means of their positions relative to the boundary and the inner boundary. As a result it turned out that points having a common predecessor and successor on both boundaries are exactly those points which can be removed from a digital set in parallel without any precautions for preventing collisions. Moreover, all simple and perfect points can be eliminated simultaneously without disturbing the topology of a set. When only neighborhood information is available (with the exception of the information whether a point is interior point or not) the set of all simple and perfect points is a maximal subset of candidates for parallel elimination (with the exception of the "dangerous" configurations of theorem 4, case 2.2).

The information whether a point is simple or not can be easily handled and updated by means of an obvious labeling. The method was implemented and run successfully with different types of binary images. Numerical results and details of implementation will be reported elsewhere.

The results presented here were obtained during a collaboration with SIEMENS Research Laboratories in München. The author greatly acknowledges stimulating discussions and valuable criticism by members of the picture processing group and specifically by G. Maderlechner and Dr. E. Hundt, the head of this group.

References

1. Eckhardt, U.: Anwendung der digitalen Topologie in der Binärbildverarbeitung. Mitteilungen der Mathematischen Gesellschaft in Hamburg, 1988 (to appear).
2. Eichhorn, B.: Skelettierung mit Randinformation. Diplomarbeit am Institut für Informatik der Technischen Universität München, Oktober 1987.
3. Lü, H. E.; Wang, P. S. P.: A comment on "A fast parallel thinning algorithm for digital patterns". Commun. of the Assoc. for Computing Machinery 29 (1986), 239-242.
4. Pavlidis, T.: Algorithms for Graphics and Image Processing. (Computer Sci. Press) Springer-Verlag, Berlin 1982.
5. Rosenfeld, A.: Digital topology. Amer. Math. Monthly 86 (1979), 621-630.
6. Tamura, H.: A comparison of line thinning algorithms from digital geometry viewpoint. Proc. of 4th Int. Conf. on Pattern Recognition Nov. 7-10, 1978, Kyoto, Japan, pp. 715-719
7. Yokoi, S.; Toriwaki, J.; Fukumura, T.: Topological properties in digitized binary pictures.Syst., Computers, Controls 4 (1973) 6, 32-40

RANDOM POINT SETS IN LATTICES

K. Voss, Jena [1)]

1. Neighborhood structures

The points of a random binary image build up black random objects in a white background or white random holes in a black background. We are interested in some topological features as functions of the only independent variable w in this theory. The value of w is the probability that any picture point is black (the colors of different points are independent from another).

The theory of random point sets in lattices is based on the theory of neighborhood structures /1/. A neighborhood structure is defined to be a tuple (P,N) where P is a finite set of points, and $N \subseteq P \times P$ is an irreflexive and symmetric relation called neighborhood relation. The set $N(p) = \{q/q \in P, (p,q) \in N\}$ is the neighborhood of point $p \in P$, and $\upsilon(p) = card(N(p))$ denotes its valence. Let $\varepsilon = card(P)$ be the number of points, and let $\varkappa = card(N)/2$ be the number of edges of a given neighborhood structure (P,N). Then, a basic formula is given by the Vertex Theorem

$$\sum_{p \in P} \upsilon(p) = 2\varkappa . \qquad (1)$$

By cyclic ordering of all points of each of the neighborhoods N(p), an orientation may be attached to a given neighborhood structure. Let C be the set of all neighborhood cycles. Then S = (P,N,C) is an oriented neighborhood structure. These neighborhood cycles coincide with a clockwise ordering of edges. The successive following of edges in respect to this clockwise ordering results in uniquely defined cyclic ways. A complete period of a generated way is called a mesh. The number of edges of a mesh m is called the lenght $\lambda(m)$ of this mesh. Any directed edge of (P,N) is contained in exactly one mesh. Let M be the set of all meshes of a given oriented neighborhood structure (shortly called as structure). Then, the Mesh Theorem

$$\sum_{m \in M} \lambda(m) = 2\varkappa \qquad (2)$$

follows. Finally, each connected structure S = (P,N,C) fullfills Euler's Theorem

$$\varepsilon - \varkappa + \mu = \chi \leq 2 . \qquad (3)$$

1) Friedrich-Schiller-University, Dept. of Technology, E. Thälmann-Ring 32, Jena, 6900, GDR

with μ = card(M) and χ as Euler's characteristic. Planar structures are characterized by a value $\chi = 2$. These three theorems may be called the topological basic equations of neighborhood structures. All further investigations are based on these three formulas.

A structure S is homogeneous if it is connected, for any point $p \in P$ we have $\upsilon(p) = \upsilon$, and $\lambda(m) = \lambda$ for any mesh of the structure. A homogeneous structure is called a net. The three basic formulas may be stated in the following form:

$$\upsilon\varepsilon = 2\varkappa, \quad \mu\lambda = 2\varkappa, \quad 2/\upsilon + 2/\lambda = 1 + \chi/\varkappa \,. \tag{4}$$

For $\chi = 2$, only the five platonic tuples $(\upsilon,\lambda,\varepsilon,\varkappa,\mu)$ = (3,3,4,6,4), (3,4,8,12,6), (3,5,20,30,12), (4,3,6,12,8), and (5,3,12,30,20) are integer solutions to (4). Also for $\chi < 0$, for any fixed value of χ there exists a finite number of small nets. Most interesting are toroidal nets with $\chi = 0$. In this case, we find always solutions for (υ, λ) = (4,4), (3,6), and (6,3). Thereby, the number ε of points can exceed any finite bound.

2. Topological quantities of regions

Assume R to be a connected region of black points in a white background with a single border mesh (which intersects the "inner" and "outer" points of the border mesh). With l as the length of the border mesh and n as the number of black-and-white connections, we obtain from (4) the equations

$$\begin{aligned} \upsilon\varepsilon - n &= 2\varkappa, \\ \lambda(\mu-1) + l &= 2\varkappa, \\ \varepsilon - \varkappa + \mu &= \chi_R. \end{aligned} \tag{5}$$

The total curvature of the border mesh is given by

$$t = n/\upsilon - l/\lambda = \chi_R - 1 \tag{6}$$

For planar subregions R of the toroidal net, t = +1 follows. Otherwise is t = -1 in the case, that R is a toroidal subregion of the toroidal net (see /1/,/2/). With f = μ-1 as number of "faces" with mesh length λ, we have at least six topological quantities for each (connected or disconnected) region:

ε number of points,	$\varkappa$ number of edges,
f number of faces,	n number of black-white connections,
t total curvature,	l length of border meshes.

It is of some interest to investigate the mean topological quantities of such regions which are building up by random distributed black points within a toroidal net.

3. Mean values

The probability of black points shall be w with $0 \leq w \leq 1$. A toroidal net with N points owns $\upsilon N/2$ edges. There are

$$\bar{x} = w^2 \cdot \upsilon N/2 \quad \text{black-black connections,}$$
$$\bar{n} = 2w(1-w) \cdot \upsilon N/2 \quad \text{black-white connections, and} \qquad (7)$$
$$(1-w)^2 \cdot \upsilon N/2 \quad \text{white-white connections.}$$

A part of all black-black connections are edges of border meshes. A directed black-black edge e is then an edge of a border mesh if at least one point of the λ points of the e-mesh is white. The probability for this situation is $1-w^{\lambda-2}$. Using $2\bar{x}$ as number of directed edges, we obtain

$$\bar{I} = (1-w^{\lambda-2}) \cdot 2\bar{x} = (1-w^{\lambda-2})w^2 \cdot \upsilon N \qquad (8)$$

as mean total length of all border meshes.

From (5) and $f = \mu - 1$ the equation

$$\bar{f} = (2\bar{x} - \bar{I})/\lambda = \upsilon N \cdot w^{\lambda}/\lambda \qquad (9)$$

follows. Finally, we obtain for the mean value of total curvature by (6) the formula

$$\bar{\tau} = \frac{\bar{n}}{\upsilon} - \frac{\bar{I}}{\lambda} = (1-w)\left[1 - \frac{\upsilon}{\lambda} \cdot \frac{w(1-w^{\lambda-2})}{1-w}\right]N. \qquad (10)$$

The following table gives some values for $\bar{\tau}/N$ as function of probability w (see /1/).

w	$\upsilon = 3$, $\lambda = 6$	$\upsilon = 4$, $\lambda = 4$	$\upsilon = 6$, $\lambda = 3$
0	0	0	0
.2	.1400	.1216	.0960
.4	.1620	.1056	.0480
.5	.1328	.0625	.0000
.6	.0833	.0096	-.0480
.7	.0238	-.0399	-.0840
.8	-.0289	-.0704	-.0960
1	-0	-0	-0

The critical probability w^* is determined by $\bar{\tau}(w^*) = 0$. We obtain

$$w^*(3,6) = .741\,270\,9105\,\ldots$$
$$w^*(4,4) = .618\,033\,9887\,\ldots$$
$$w^*(6,3) = .5$$

with $w^*(6,3) = 1/2$ as exact value.

The physicists know similar numerical values investigating percolation problems in lattices /3,4,5/. Only approximative results are found by computer simulation experiments for large planar lattices $(\nu, \lambda) = (3,6)$ and $(\nu, \lambda) = (4,4)$. The values are $\tilde{w}(3,6) \approx 0.70$ and $\tilde{w}(4,4) \approx 0.59275 \pm 0.00003$. There arise the problem wether for toroidal nets the relation $w^* = \tilde{w}$ is right.

4. Literatur

/1/ Voss, K.: Theoretische Grundlagen der digitalen Bildverarbeitung. Akademie-Verlag, Berlin 1988

/2/ Klette, R.; K. Voss: The tree basic formulas of neighborhood structures. In: L. P. Yaroslavskij, A. Rosenfeld, W. Wilhelmi (eds).: Computer Analysis of Images and Patterns. Mathem. Res. Vol. 40, Akademie-Verlag, Berlin 1987, p. 149-161.

/3/ Seymor, P. D.; D. J. A. Welsh: Percolation probabilities on the square lattices. Ann. Discrete Math. 3 (1978) 227-245.

/4/ Essam, J. W.: Percolation theory. Reports on Progress in Physics. 43(1980) 833-912.

/5/ Kesten, H.: Percolation Theory for Mathematicians. Birkhäuser, Boston-Basel-Stuttgart 1982

MOTIONS IN THE DISCRETE PLANE

by Albrecht Hübler [1)]

1. Introduction

The traditional way to study geometrical problems over several grids (We omit references concerning this area because the main ideas, methods, results, and important authors as well, can be assumed to be well known.) by "translation" of the usual Euclidean geometry into the underlying discrete structure according to a fixed "digitization mapping" has been useful for solving elementary practical problems of geometry in image processing, and for handling special problems which are closely related to the quantization process. But, this approach of the "digital geometry" has not been suited to bring forth a selfcontained mathematical theory of discrete geometry for image processing. Nevertheless, there is a need for a discrete theory based on discrete models which reflects the discrete nature of digital images in a correct manner.

For about five years several scientists in Berlin and Jena concentrate their investigations to special areas of mathematical-theoretical foundation of image processing. The goal is to develop general models and theories to describe image structures and corresponding discrete topology and geometry in a way which is abstract enough to be free of technical details, but which preserves essential and typical aspects of supporting domains of digital images. Results can be found in the given references.

2. An axiomatic theory of discrete geometry

In [1,2] was presented an axiomatic approach to a discrete image geometry. The notion of translations and its relations to parallelity and betweenness relation for point triples play a central role. This theory represents a very general approach which allows to study discrete geometry with respect to several kinds of grids or neighborhood structures (see [4],[8]-[10] for this notion) based on the same theoretical foundation. Important notions, like straight lines and convexity, are closely related to the corresponding notions in the traditional digital geometry [1],thus providing the relevance of our axiomatic approach for an image processing theory. It is an important advantage that, formulating the axioms, basic properties and also essential differences from the usual Euclidean geometry are expressed explicitely.

[1)] Friedrich-Schiller-Universität, Sektion Mathematik
UHH, 17.OG, JENA, DDR-6900

The present paper discusses an approach to continue this theory by introduction of important geometrical transformations like reflections and rotations, thus being able to describe geometrical motions in the discrete plane. Before we can present these new results, it is necessary to give a short overview about the essential ideas and properties of the axiomatic theory. For raesons of space limit all results can be presented without proofs only. Let P be a set of points and G a nonempty set of subsets of P, the set of straight lines, given as basic objects:

Axioms of incidence:

AXIOM 1 : For any two different points p and q there is exactly one straight line g(p,q) containing p and q.

AXIOM 2 : For any straight line g there exist three pairwise different points p,q, and r with $p,q \in g$ and $r \notin g$.

AXIOM 3 : On the set G an equivalence relation $||$ is fixed, such that for any straight line g and an arbitrary point p there is exactly one straight line g' with $p \in g$ and $g \,||\, g'$.

DEFINITION : A bijective mapping A on P is called a translation, if it is either the identity ID or it fulfils the following properties:

(T-1) For $g \in G$ is $A(g) := \{ A(p) : p \in g \}$ a straight line which is parallel to g.

(T-2) For all $p \in g$ holds $A(p) \neq p$.

(T-3) The set of straight lines $\{ g(p,A(p)) : p \in P \}$ is an equivalence class with respect to $||$.

AXIOM 4 : For points $p,q \in P$ there exists a translation A with A(p)=q.

Axioms of ordering:

As a further kind of basic objects of the theory let be fixed two order relations $<$ and $>$ on each straight line g, such that for points p,q of g the relation $p < q$ holds, if and only if $q > p$ is true. We call the pairs [g,<] and [g,>] oriented straight lines.

DEFINITION : For three pairwise different points p,q,r of an oriented straight line [g,<] the point q lies between p and r (We shortly write Z(p,q,r)) , if and only if $p < q < r$ or $p > q > r$ is true.

AXIOM 5 : For any oriented straight line [g,<] and arbitrary $p \in g$ there exist points q and r of g , such that Z(q,p,r).

AXIOM 6 : For any two different points p,q there is at most a finite number of points r with Z(p,r,q).
(First axiom of discreteness)

AXIOM 7 : Let be g1,g2 and g3 pairwise different parallel straight lines, and g and g' are straight lines which have common points with each of the straight lines g1,g2,g3. Then for $\{ p_i \} = g \cap gi$, and for

$\{ p_i' \} = g' \wedge gi$, $i = 1,2,3$, the relation $Z(p_1, p_2, p_3)$ is true, if and only if $Z(p_1', p_2', p_3')$ holds.

COROLLARIES :

(1) For any straight line g there is a translation A, such that g can be represented as $g = \{ A^i(p) : i \in \Gamma \}$, where p is an arbitrary point of g. (Γ denotes the set of integers, and A means the i-fold concatenation of A for $i > 0$, the identity for $i = 0$, and the $(-i)$-fold concatenation of the inverse A^{-1} of A for $i < 0$.)
We call A a generator of the straight line g.

(2) For any straight line g there are exactly two generators A and A^{-1}.

(3) The following statements are equivalent for arbitrary straight lines g and g':
- g and g' are parallel;
- There is a translation $A \neq ID$, such that $A(g) = g$ and $A(g') = g'$;
- There is a translation B with $B(g) = g'$;
- g and g' have the same generators.

(4) Any straight line has exactly enumerable infinite many points.

(5) The betweenness relation for points is invariant with respect to translations, i.e. for a translation A and points p,q,r with $Z(p,q,r)$ also $Z(A(p),A(q),A(r))$ holds.

DEFINITION : For three parallel straight lines $g1,g2$, and $g3$ we say that $g2$ lies between $g1$ and $g3$, if there exist a translation $A = ID$ and positive integers i and j, such that $A^i(g1) = g2$ and $A^j(g2) = g3$. Because there is no danger of confusion we use the same symbolic abbreviation as for the point-betweenness relation: We write $Z(g1,g2,g3)$.

DEFINITION : A subset P' of P is called planar, if P' contains triples of non-colinear points, and for any four points p,q,r,s of P' holds: For the straight lines $g1,g2,g3$ defined by $g1 = g(p,q)$, $g2 \parallel g1$ and $r \in g2$, $g3 \parallel g1$ and $s \in g3$, follows $Z(g1,g2,g3)$ or $Z(g1,g3,g2)$ or $Z(g2,g1,g3)$, if $g1,g2$, and $g3$ are pairwise different straight lines.

DEFINITION : A subset P' of P is called a plane in P, if P' is a maximal planar point set, i.e. if for any point $p \in P \setminus P'$ the set $P' \cup \{ p \}$ is not planar.

DEFINITION : A planar subset P' of P is called a planar grid, if there are translations A and B, such that for an arbitrary point p of P' the set P' is representable by $P' = \{ A^i \circ B^j(p) : i,j \in \Gamma \}$, where $\circ$ is the operation of concatenation. We say that the translations A and B generate P', or in an equivalent manner, the straight lines $g1 = \{ A^i(p) : i \in \Gamma \}$ and $g2 = \{ B(p) : i \in \Gamma \}$ span P'.

AXIOM 8 : For any two different parallel straight lines $g1$ and $g2$ there exists at most a finite number of parallel straight lines $g3$ with $Z(g1,g2,g3)$.
(Second axiom of discreteness)

COROLLARY : Any plane in P is a planar grid.
(Remark that this corollary is not true without axiom 8.)

3. Discrete planes and coordinate systems

All considerations in this and in the following sections are for discrete planes. Any discrete plane P is a planar grid, hence two straight lines g1 and g2 exist which span P. Let 0 be the common point of g1 and g2, and A and B generators of g1 and g2 , respectively. An arbitrary point p of P is determined uniquely by two integers x and y with $p = A^x \circ B^y(0)$. We call the pair (g1,g2) a coordinate system of P with the origin 0. The integers x = x(p) and y = y(p) are the coordinates of p with respect to (g1,g2).
Let us in our further considerations refer to a fix reference coordinate system (gx,gy). Let X and Y be generators of gx and gy , respectively, and x = x(p) and y = y(p) the coordinate functions with respect to (gx,gy).
We consider two straight lines g1 and g2. Let A and B be two fixed generators of g1 and g2, respectively. We can represent A and B by the coordinates x1 = x(A(0)), y1 = y(A(0)), x2 = x(B(0)), y2 = y(B(0)). Thus a coordinate system (g1,g2) can be described uniquely by its origin and by the matrix

$$M(g1,g2) = \begin{pmatrix} x1 & x2 \\ y1 & y2 \end{pmatrix}.$$

For any ordered triple (p,q,r) of points we say that (p,q,r) is positively (negatively) oriented, if the value

$$Or(p,q,r) = \det \begin{pmatrix} 1 & x(p) & y(p) \\ 1 & x(q) & y(q) \\ 1 & x(r) & y(r) \end{pmatrix}$$ is greater (less) than 0.

For a coordinate system (g1,g2) its orientation is given by Or(0,A(0),B(0)) , where A and B are the fixed generators of g1 and g2, respectively.

COROLLARY : A coordinate system (g1,g2) is positively (negatively) oriented, if and only if det M(g1,g2) = 1 (det M(g1,g2) = -1).

4. Discrete geometrical mappings

DEFINITION : A bijective mapping F from a discrete plane P to P is called a geometrical mapping, if the betweenness relation for points is invariant with respect to F, i.e. if for any three points p,q,r of P from Z(p,q,r) follows Z(F(p),F(q),F(r)).

COROLLARY : Translations are geometrical mappings.

THEOREM : A bijective mapping F from P to P is a geometrical mapping, if and only if for the coordinate system (gx,gy) holds that (F(gx),F(gy)) is a coordinate system again, and for any point p of P its image F(p) has with respect to the coordinate system (F(gx),F(gy)) the same coordinates like p with respect to (gx,gy).

With the last theorem we can speak about the class of geometrical mappings as of the class of bijective coordinate transformations in the discrete plane as well.

Like in the usual analytical geometry, those discrete geometrical mappings F which let the origin fix, can be represented by a transformation matrix

$$MF = \begin{pmatrix} x1 & x2 \\ y1 & y2 \end{pmatrix},$$ such that for a point p the equation

$$\begin{pmatrix} x(F(p)) \\ y(F(p)) \end{pmatrix} = MF \cdot \begin{pmatrix} x(p) \\ y(p) \end{pmatrix}$$ is true.

Let (g1,g2) be the coordinate system which is the F-image of (gx,gy). It follows M(g1,g2) = MF immediately. Thus we have det MF = 1 or det MF = -1.

An arbitrary discrete geometrical mapping can be represented as the concatenation of an origin-preserving geometrical mapping and a translation.

5. Motions in the discrete plane

Our interest is now directed to special kinds of discrete geometrical mappings. We want to find out how to define geometrical motions. A priori the class TR of translations should belong to the class of motions. To get the classes of discrete rotations and discrete reflections we use the following approach:

DEFINITION : A generalized discrete rotation R (with rotation center 0) is an origin-preserving geometrical mapping with the properties

(R1) R is the identity or 0 is the only one fixpoint of R.

(R2) det MR = 1 (Orientation of coordinate system is invariant).

(R3) If

$$MR = \begin{pmatrix} x1 & x2 \\ y1 & y2 \end{pmatrix}$$,then $x2 = y1$ or $x2 \cdot y1 < 0$, and $x1 \cdot y2 > 0$.

(Orientations of new x-axis, with respect to the old x-axis and to the old y-axis, are the same as the orientations of the new y-axis, with respect to the old y-axis and to the old x-axis, respectively.)

THEOREM : An origin-preserving geometrical mapping R is in the set ROT of generalized discrete rotations, if and only if its transformation matrix MR is in the set

$$\left\{\begin{pmatrix} a & 1 \\ -1 & 0 \end{pmatrix}, \begin{pmatrix} 0 & 1 \\ -1 & a \end{pmatrix}, \begin{pmatrix} a & -1 \\ 1 & 0 \end{pmatrix}, \begin{pmatrix} 0 & -1 \\ 1 & a \end{pmatrix}, \begin{pmatrix} 1 & 0 \\ 0 & 1 \end{pmatrix}, \begin{pmatrix} -1 & 0 \\ 0 & -1 \end{pmatrix} : a \in \Gamma\right\}$$

It is easy to prove that ROT is not closed under concatenation. There are exactly three different maximal cyclic groups formed by elements of ROT:

$$ROT1 = \{ R1 \ : i = 1,\ldots,4 \} \quad \text{with} \quad MR1 = \begin{pmatrix} 0 & -1 \\ 1 & 0 \end{pmatrix},$$

$$ROT2 = \{ R2 \ : i = 1,\ldots,6 \} \quad \text{with} \quad MR2 = \begin{pmatrix} 1 & -1 \\ 1 & 0 \end{pmatrix},$$

$$ROT3 = \{ R3 \ : i = 1,\ldots,6 \} \quad \text{with} \quad MR3 = \begin{pmatrix} 0 & -1 \\ 1 & 1 \end{pmatrix}.$$

DEFINITION : A generalized (origin-preserving) discrete reflection S is an origin-preserving geometrical mapping with the properties:

(S1) There is exactly one fixpoint-straight line g with respect to S.

(S2) det MS = -1 (Orientation of coordinate system changes).

(S3) If

$$MS = \begin{pmatrix} x1 & x2 \\ y1 & y2 \end{pmatrix} \quad \text{then} \quad x2\ y1 > 0 \quad \text{and} \quad x1\ y2 \leq 0.$$

(Orientations of new x-axis with respect to the old x-axis and to the old y-axis are different from the orientations of the new y-axis with respect to the old y-axis and the old x-axis, respectively.)

THEOREM : An origin-preserving geometrical mapping S is in the set REF of generalized discrete reflections, if and only if its transformation matrix MS is in the set

$$\left\{\begin{pmatrix} 0 & 1 \\ 1 & 0 \end{pmatrix}, \begin{pmatrix} 0 & -1 \\ -1 & 0 \end{pmatrix}, \begin{pmatrix} -1 & a \\ 0 & 1 \end{pmatrix}, \begin{pmatrix} 1 & a \\ 0 & -1 \end{pmatrix}, \begin{pmatrix} -1 & 0 \\ a & 1 \end{pmatrix}, \begin{pmatrix} 1 & 0 \\ a & -1 \end{pmatrix} : a \in \Gamma, a \neq 0\right\}.$$

If we now demand that with respect to the reference-coordinate system the motions can be represented as concatenations of elements of ROT, REF ,and TR, and that the class MOT of motions is closed under concatenation, and maximal (with respect to the set theoretical inclusion), we get three different maximal motion groups:

THEOREM : With respect to a fixed reference-coordinate system the class MOT can be defined to be one of the following three transformation classes:

MOT1 = (ROT1 $\cup$ REF1) $\circ$ TR with REF1 = { S1,S2,S3,S4 }, where

$$MS1 = \begin{pmatrix} 0 & 1 \\ 1 & 0 \end{pmatrix} \quad MS2 = \begin{pmatrix} 0 & -1 \\ -1 & 0 \end{pmatrix} \quad MS3 = \begin{pmatrix} -1 & 0 \\ 0 & 1 \end{pmatrix} \quad MS4 = \begin{pmatrix} 1 & 0 \\ 0 & -1 \end{pmatrix};$$

MOT2 = (ROT2 $\cup$ REF2) $\circ$ TR with REF2 = { S1,S2,S5,S6,S7,S8 }, where

$$MS5 = \begin{pmatrix} -1 & 0 \\ -1 & 1 \end{pmatrix} \quad MS6 = \begin{pmatrix} 1 & 0 \\ 1 & -1 \end{pmatrix} \quad MS7 = \begin{pmatrix} 1 & -1 \\ 0 & -1 \end{pmatrix} \quad MS8 = \begin{pmatrix} -1 & 1 \\ 0 & 1 \end{pmatrix};$$

MOT3 = (ROT3 $\cup$ REF3) $\circ$ TR with REF3 = { S1,S2,S9,S10,S11,S12 }, where

$$MS9= \begin{pmatrix} -1 & 0 \\ 1 & 1 \end{pmatrix} \quad MS10= \begin{pmatrix} 1 & 0 \\ -1 & -1 \end{pmatrix} \quad MS11= \begin{pmatrix} 1 & 1 \\ 0 & -1 \end{pmatrix} \quad MS12= \begin{pmatrix} -1 & -1 \\ 0 & 1 \end{pmatrix}$$

The last theorem can be interpreted in the following way:
Choosing the class MOT1 as the class of motions, corresponds to the interpretation of the discrete plane as the 4-neighborhood graph with four origin-preserving rotations and four origin-preserving reflections (with reflection-axis x=0, y=0, y=x, and y=-x, respectively).
Choosing the class MOT2 or MOT3 as the class of motions, respectively, corresponds to the interpretation of the discrete plane as the planar 6-neighborhood graph with six origin-preserving rotations and six origin-preserving reflections (with reflection-axis determined by the 3 main directions in the 6-neighborhood graph and by those straight lines which cut the sectors between the main directions in halfs). The only difference between MOT2 and MOT3 is the different choice of x- and y-axis between the three possible main directions in 6-neighborhood graphs.
The following figures illustrate these interpretations for MOT2 and MOT3.

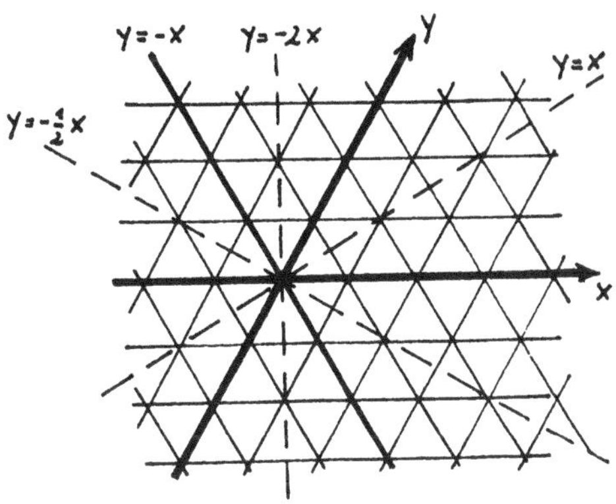

Fig. 1:
6-neighborhood graph with CG3-interpretation

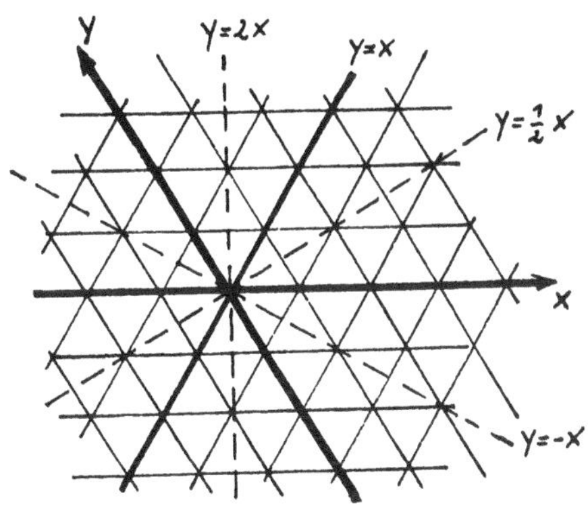

Fig. 2:
6-neighborhood graph with CG2-interpretation

6. References

[1] Hübler, A.: An axiomatic approach to discrete geometry and its relation to usual digital geometry for image processing. In: L.P.Yaroslavskij, A.Rosenfeld, W.Wilhelmi (eds.): Computer Analysis of Images and Patterns. Math.Research 40, Berlin, Akademie-Verlag 1987, 174-186

[2] Hübler, A., R. Klette : Theoretische Grundlagen der digitalen Bildverarbeitung, VII. Geometrie auf Nachbarschaftsstrukturen. Bild und Ton 40 (1987)11, 229-235

[3] Hübler, A. : A general conception of discrete straight lines and convexity. In: I.Plander (edt.): Artificial Intelligence and Information-Control Systems of Robots-87, North Holland, 1987

[4] Klette, R., K. Voss : The three basic formulae of neighbourhood structures. In: L.P.Yaroslavskij, A.Rosenfeld, W.Wilhelmi (eds.): Computer Analysis of Images and Patterns. Mathem.Research 40, Berlin, Akademie-Verlag 1987, 149-161

[5] Kovalevskij, V.A.: On the topology of discrete spaces. Tagungsberichte Digitale Bildverarbeitung, TU Dresden, April 1986,56-77

[6] Kovalevskij, V.A.: The topology of cellular complexes as applied to image processing. In: L.P.Yaroslavskij, A.Rosenfeld, W.Wilhelmi (eds.): Computer Analysis of Images and Patterns. Mathem. Research 40, Berlin, Akademie-Verlag 1987, 162-173

[7] Voss, K. : Discrete integral geometry. In: L.P.Yaroslavskij, A.Rosenfeld, W.Wilhelmi (eds.): Computer Analysis of Images and Patterns. Mathem.Research 40, Berlin, Akademie-Verlag 1987,209-212

[8] Voss, K. : Nachbarschaftsstrukturen als Bildtraeger. In: E.Paulus (Hrsg.): Mustererkennung 1987, Proc. 9.DAGM-Symposium, Braunschweig 1987, Informatik-Fachberichte 149, Berlin Heidelberg, Springer 1987 141-150

[9] Voss, K. : Theoretische Grundlagen der digitalen Bildverarbeitung. Berlin, Akademie-Verlag 1988

[10] Voss, K., P. Hufnagl, R. Klette : Theoretische Grundlagen der digitalen Bildverarbeitung, I. Bild und Ton 38 (1985) 299-302 and subsequent papers in Bild und Ton 38(1985)325-331,39(1986)54-50, 213-219,303-307,40(1987)112-118

SOME PROPERTIES OF THE UNIMODULAR GROUP IN R^2

Michael Bitter [1)]

1. Introduction

The examination of properties of objects is a fundamental task in automatic image analysis. Since a computer can store a finite number of informations only, it is necessary to map the real space onto a discrete structúre such that the interesting objects have an image consisting of a finite number of elements. Such a mapping is said to be a digitization. The orthogonal grid is an often used discrete structure, wich we can identify with the integer lattice Z^d in the d-dimensional Euclidean space R^d.

Quantities such as volume, surface, length or the Euler-Poincare characteristic are basic properties for real objects. All mentioned quantities are able to be expressed as Quermass integrals. For the study of these functionals a rich theory has been developed. However their determination raises objections. One problem is that many informations about the real object are lost by digitization. A second problem is given by the fact, that it is difficult to use stereological or other methods in the integer lattice, because the known methods are based on qualities of the Euclidean space and the Euclidean motions.

Since the lost of information by digitization is unavoidable, in several cases it can be sensible to assume, that the real objects are bounded polyhedrons with integer vertices only, called lattice polyhedrons.

For the set P^d of lattice polyhedrons in R^d among the Quermass integrals there exists another class of functionals with interesting properties. That class, called lattice point functionals, was found by Ehrhart.[4] Many analogies for Quermass integrals and lattice point functionals were found. That is obvios by the next.[1][2][3][4]

For P^d there are d+1 lattice point functionals G_0, $G_1, \dots, G_d$ with the following properties. Any functional G_i is additive, unimodular invariant and homogeneous of degree i. The conceptions of additivity and homogenity âre the same as for Quermass integrals. The unimodular invariance refers to the group of those affine transformations in R^d, whose restrictions on the integer lattice are one to one transformations on the same.

A fundamental characterization of the lattice point functionals was given by Betke: The set of the additive and unimodular invariant functionals on P^d is a (d+1)-dimensional real vektor space with the lattice point functionals $G_0, \dots, G_d$ as a base.

1) Michael Bitter Friedrich-Schiller-Universität Sektion Mathematik UHH 17. OG DDR-6900 Jena

That result is an analogy to Hadwiger's characterization of the Quermass integrals.

This time there are many papers, which are engaged with the lattice point functionals. But in difference to the Quermass integrals there is a little geometric interpretation for the G_i only. It is known, that G_d is the volume and G_0 is the Euler-Poincare characteristic. For G_{d-1} a geometric interpretation is given also. However the significance of the other G_i is unknown.

We think, that a greater sense for the complex structure of the integer lattice and the unimodular transformations could give new chances to get novel expressions for the lattice point functionals, wich can manage interpretations of these functionals as topological properties of objects of P^d

In automatic image analysis such interpretations could offer new possibilities to solve given problems.

In this sense the present work is to be a first small step. It relates to the case of dimension 2 only.

2. The unimodular group in R^2 and it's properties

Let R^d and Z denote the d-dimensional Euklidean space and the set of all integers, respectively.

Definition 1: A mapping $f: R^d \rightarrow R^d$ is said to be an unimodular transformation if and only if $f(x) = Ax + b$, $A = (a_{ij})_{i,j=1}^{d}$, $b = (b_1, \ldots, b_d)$, $a_{ij} \in Z$, $b_i \in Z$, $\det A = \pm 1$. [1]

To express the concatenation of two unimodular transformations A and B we write A*B. M^d denotes the group of all unimodular transformations with operation '*' and identity E as unit element. Further let L^d be the subgroup of linear transformations in M^d.

Lemma 1: Let $A = \begin{pmatrix} a & b \\ c & d \end{pmatrix}$ be an element of L^2 with $A \neq \pm E$. The cyclic group g(A) generated by A is finite if and only if $0 \leqslant |a+d| \cdot \det A \leqslant 1$. For $\det A = 1$ the group g(A) is of order 3, 4, 6 for $a+d = -1, 0, 1$, respectively.

In the case of $\det A = -1$ and $a+d = 0$ the order of g(A) is 2.

To prove this statement one can use the fact, that for $\det A = 1$ A satisfies it's charcteristic equation

$$0 = A^2 - (a+d)A + E,$$

and by induction we obtain

$$A^k = P_k(a+d)A - P_{k-1}(a+d)E$$

with $P_0(x) \equiv 0$, $P_1(x) \equiv 1$, $P_{k+1}(x) = xP_k(x) - P_{k-1}(x)$.

The polynomials $P_k(x)$ are related to the Tschebyscheff polynomials of second kind. [5] By examining the points of zero and using the fact that $a+d \in Z$ we obtain the statement of Lemma 1 for $\det A = 1$.
For $\det A = -1$ we have $A^2 = (a+d)A + E$, and from $a+d = 0$ the identity $A^2 = E$ follows. The case $a+d \neq 0$ can be reduced to the case $\det A = 1$.

For a, b, c, d $\in R^1$ any order of the cyclic group $g(A)$ is possible. The requirement of a, b, c, d to be integers is the reason for the limitation of the order of cyclic groups and gets the restriction of A to Z to be a bijection of Z^2 to Z^2.
Referring to Lemma 1 one can regard the cyclic groups of finite order 3, 4, 6 and the group generated by $-E$ as generalization of the cyclic groups of rotations in the Euklidean plane with order 3, 4, 6 and 2, respectively.
In this sense for Z^2 the finite cyclic subgroups, generated by elements A of L^2 with $\det A = 1$, are an analogy to the cyclic groups of order 2, 3, 4, 6 generated by rotations in R^2.
The same analogy exists between the groups, generated by an A with $\det A = -1$, $A^2 = E$, and reflections in R^2.

Now we regard the following cyclic transformations of L^2, which are examples for generators of cyclic groups of order 6, 4 and 2. Let

$$D_6 = \begin{pmatrix} 1 & -1 \\ 1 & 0 \end{pmatrix}, \quad D_4 = \begin{pmatrix} 0 & -1 \\ 1 & 0 \end{pmatrix}, \quad S_2 = \begin{pmatrix} 0 & 1 \\ 1 & 0 \end{pmatrix} \quad \text{and } C = \{D_6, D_4, S_2\}.$$

Theorem 1: Any Transformation of L^2 can be obtained as a concatenation of a finite number of elements of C.
This means, that for any $A \in L^2$ there is a sequence $A_1, A_2, \dots, A_t$ with $A_i \in C$, such that $A = A_1 * A_2 * \dots * A_t$.

Proof: Let $A = \begin{pmatrix} a & b \\ c & d \end{pmatrix} \in L^2$ with $\det A = 1$.

First case: Let $a = 0$.
From $a = 0$ follows $A = \begin{pmatrix} 0 & -j \\ j & k \end{pmatrix}$ with $j = \pm 1$, $k \in Z$.

It is easy to verify, that

$$A = D_4^j * (D_6 * D_4)^{kj}, \quad j = \pm 1$$

and so for the first case the proof is complete.

Second case: Let $b = 0$.
If $b = 0$ we obtain by $A = B * D_4$ with $B = \begin{pmatrix} b & -a \\ d & -c \end{pmatrix}$ the situation of the first case.

Third case: Let $a \neq 0$ and $b \neq 0$.
Now we consider the mapping $f: L^2 \to L^2$ with

$$f(A) = \begin{cases} A*D_6 = \begin{pmatrix} a+b & -a \\ c+d & -c \end{pmatrix} & \text{for } \mathrm{sgn}(a) \neq \mathrm{sgn}(b) \text{ and } |a| \leq |b| \\ A*D_6^2 = \begin{pmatrix} b & -a-b \\ d & -c-d \end{pmatrix} & \text{for } \mathrm{sgn}(a) \neq \mathrm{sgn}(b) \text{ and } |a| > |b| \\ A*D_4*D_6*D_4^3 = \begin{pmatrix} a-b & a \\ c-d & c \end{pmatrix} & \text{for } \mathrm{sgn}(a) = \mathrm{sgn}(b) \text{ and } |a| \leq |b| \\ A*D_4*D_6^4*D_4^3 = \begin{pmatrix} -b & a-b \\ -d & c-d \end{pmatrix} & \text{for } \mathrm{sgn}(a) = \mathrm{sgn}(b) \text{ and } |a| > |b| \end{cases}$$

Here $\mathrm{sgn}(x) = x/|x|$.
Now let

$$A_k = \begin{pmatrix} a_k & b_k \\ c_k & d_k \end{pmatrix} = f^k(A), \quad k \in Z.$$

Then for $a_k \neq 0$ and $b_k \neq 0$ we obtain

$$|a_{k+1}| + |b_{k+1}| < |a_k| + |b_k| .$$

Thus there exists a $n \in Z$ with $a_n = 0$ or $b_n = 0$, and we have $A*B_1*...*B_s = A_n$ with $a_n = 0$ or $b_n = 0$ and $B_k \in C$, $k = 1...s$.
Applying the first or the second case for A_n and using the equations

$$A = A_n*B_s^{-1}*...*B_1^{-1}, \quad D_i^{-1} = D_i^{i-1}, \quad i = 4,\ 6,$$

we obtain the statement of the theorem for $\det A = 1$.

If $\det A = -1$ we note, that $A = S*S*A$.
Since $\det S*A = 1$ we can apply the cases 1, 2 or 3 to $S*A$ and now the proof is complete.
It is evident, that any translation in M^2 can be constituted by multiple application of two simple translations T_1 and T_2 with translation vectors (0,1) and (1,0), respectively. This fact and Theorem 1 yield the following result.

Corollary 1: Any transformation of M^2 can be obtained by concatenation of a finite number of elements from the set $\{D_6, D_4, S_2, T_1, T_2\}$.

3. An Application

Let P^2 be the set of all lattice polytops in R^2. [1]

Definition 2: A mapping $h: P^2 \to R^1$ is called unimodular invariant if and only if $h(p) = h(B(p))$ for all $p \in P^2$ and all $B \in M^2$. [1]
Here $h(p)$ denotes the image of p by applying h.

Theorem 2: A mapping $h: P^2 \rightarrow R^1$ is unimodular invariant if and only if $h(p) = h(B(p))$ for all $p \in P^2$ and all $B \in \{D_6, D_4, S_2, T_1, T_2\}$.

The proof of this theorem follows as a consequence of Corollary 1.

Hence the check for unimodular invariance of a mapping h is reduced to check the five transformations of the set $\{D_6, D_4, S_2, T_1, T_2\}$ only.
A construction of additive and unimodular invariant transformations by using this fact possibly gets new geometric interpretations for the lattice point functionals.

References

[1] Betke, U.: Gitterpunkte und Gitterpunktfunktionale, Habilitationsschrift, Siegen 1979

[2] Betke, U., McMullen, P.: Lattice Points in Lattice Polytopes Monatshefte für Mathematik, 99 (1985), 253-265

[3] Betke, U., Kneser, M.: Zerlegungen und Bewertungen von Gitterpolytopen, Journal für die reine und angewandte Mathematik, 358 (1985), 202-208

[4] Ehrhart, E.: Sur un probleme de geometrie diophantienne lineaire, Journal für die reine und angewandte Mathematik, 226 (1967), 1-29 und 227 (1967), 25-49

[5] Obreschkoff, N.: Verteilung und Berechnung der Nullstellen reeller Polynome, Deutscher Verlag der Wissenschaften, Berlin 1963

OPTIMAL LINES IN THE PLANE: A SURVEY

Korneenko N.M.[1)]

1. Introduction. This brief survey deals with various problems of locating a straight line with respect to some planar configuration (scene), e.g. a finite pointset or a collection of polygons, etc. The notion of "optimal line" will be rather vague: sometimes we only mean to find a line with some desirable property w.r.t. the scene, e.g. intersecting all objects of the scene. The main concern of this paper is to demonstrate quite a few results obtained by methods of Computational Geometry in its new spirit of [19, 30, 10]. We will proceed by giving first a formulation of a problem, then a short discussion goes with computational complexity results. Perhaps, to understand some comments the reader must address himself to original papers.

Notation: $P = \{p_1,..,p_n\}$ - a planar pointset, $p_i=(x_i,y_i)$; points may have weights $w_i \geqslant 0$. d_E - Euclidean distance; $d_M(p_i,p_j)$ - Manhattan distance (equals to $|x_i-x_j|+|y_i-y_j|$). L is a line in the plane. For a problem tagged "XNN" L_{XNN} is the desired (optimal) line.

2. Regression-like lines. Scene: a finite planar pointset with or without weights. We begin with paying our respect to classic least-squares lines.

Problems R1, R2, R3 (R is for regression). Minimize the sum of squared 1) vertical, 2) horizontal, or 3) normal to a line deviations of P from a straight line (both weighted and unweighted cases).

The smoothness of corresponding cost functions gives rise to closed-form analytic formulae evaluated in linear time. However in some cases the LSM can result in serious errors [22]. So one might be interested in other criteria for the best linear fit.

Problem R4 (minsum). $\sum_i w_i d_E(p_i,L) \to \min_L$

Here the cost function is no longer smooth and the problem gains a combinatorial flavor. In [26] a discrete characterization is given for optimal lines L_{R4}: any line L_{R4} passes through some pair of points p_i - hence the straightforward $O(n^3)$-time algorithm. The second property: sums of weights of points on the either side of L_{R4} do not exceed the halfsum of all weights - was used in [26] to prune away some cost function computations without complexity estimates. In [2, 3, 25] these properties together with some angular sorting have led to $O(n^2 \log n)$-time $O(n)$-space algorithms. (Originally [25] needed $O(n^2)$ space to sort all pair-of-points lines by slope, but it can be avoided [2, 3, 18]).

1) USSR 220604 Minsk-72, Surganova 11.
Institute of mathematics. Byelorussian academy of sciences.

1N 18 an $O(n^2)$-time and space algorithm was proposed based on a method of simultaneous sorting of P by polar angles around each p_i from [7]. Here a renowned duality transform between points and nonvertical lines of the plane (polarity) plays crucial role. It is useful for many other problems involving lines (see below) so we put it down here.

$$T: \quad p=(a,b) \rightarrow T_p:(y=ax+b)$$
$$L:(y=kx+d) \rightarrow T_L=(-k,d)$$

For the weighted case the two properties of L_{R4} did their best: in [2] an example of P is given with $O(n^2)$ lines posessing these properties so the complexity lower bound is Omega(n^2). However in the unweighted case (all w_i are equal) the number h(n) of these "halving lines" is bounded from the above by $O(n^{3/2})$ [21, 13] and it is conjectured that this ded is not tight. We give a sketch of a new $o(n^2)$-time algorithm for the unweighted case. We use a halving line rotation procedure due to Lovasz [21, 13]. Assume first that no three points are on the same line. Starting from an arbitrary halving line (which can be found in O(nlogn) time) passing through points, say, p_1 and p_2 we rotate it clockwise until it hits some p_3. Then rotate it around p_3 and so on. Every line (p_{2k-1},p_{2k}) will be halving and all halving lines will be encountered. For each pair of these consecutive lines the cost function can be recomputed in O(1) time [3, 25]. In order to find next line during the rotation we may dynamically maintain convex hulls of the left and right pointsets and search tangents to them in time $O(\log^2 n)$ per line with O(nlogn) preprocessing [29]. So we end with an $O(n^{3/2}\log^2 n)$-time algorithm (more precisely, $O(n\log n+h(n)\log^2 n)$-time). When we drop away the three-point constraint the cost function updating is no longer O(1)-time (there are also some subtleties in the proper definition of the rotation process). In this case we must take into account the number of point-line incidences between P and the halving line set: according to [33] its worst-case estimate is $O(n^{2/3}h(n)^{2/3}) = O(n^{5/3})$ and it dominates the time complexity.

<u>Open problem</u> <u>O1</u>. Make a sharper estimate of this halving line rotation procedure and the combinatorial quantities involved.

<u>Problem</u> <u>R5</u> [25]. $\sum_i w_i d_M(p_i,L) \rightarrow \min_L$

It turns out that in this rather strange use of Manhattan metric the point-line distance is measured parallel to the y-axis when the line line slope is between +1 and -1 and parallel to the x-axis otherwise. So R5 splits into R6 and R7.

<u>Problem</u> <u>R6</u>. $\sum_i w_i |y_i-ax_i-b| \rightarrow \min_{a,b}$

<u>Problem</u> <u>R7</u>. $\sum_i w_i |x_i-ay_i-b| \rightarrow \min_{a,b}$

For them in [25] an $O(n \log^2 n)$-time rather involved algorithm is given. Note that R6,7 are also quite common in mathematical statistics and people say that they can be solved by linear programming. What about applying Megiddo's prune-and-search to them?

Problem R8 (minmax). $\max_i w_i d_E(p_i,L) \rightarrow \min_L$

In the unweighted case it is equivalent to the "width" problem R8*.

Problem R8*. For a planar pointset find a minimum width strip between two parallel lines containing it.

It was solved repeatedly in [1, 9, 16, 20] (and could be obtained as a byproduct in [8]) in O(nlogn) time by basically the same procedure of rotating calipers. [20] shows that it is optimal, i.e. a lower bound of Omega(nlogn) is proven.

For the weighted case [27] gives a discrete characterization of L_{R8} leading to a straightforward $O(n^3)$-time algorithm: L_{R8} lies on eaqual (weighted) distances from at least three different farthest points. In [2, 3] and [20] different $O(n^2 \log n)$-time, $O(n^2)$-space algorithms are given. In [20] the bottleneck is the computation of the convex hull of $O(n^2)$ so-called "weighted composite points". In [2, 3] the bottlenecks are the computation of n convex hulls of groups of neighbouring "equilibrum points" and the repeated test if an "equilibrum point" is an extremeof a given convex hull (O(logn) per test by binary search). The binary searches can be replaced by an O(1)-time retrieval from an auxiliary table of size $O(n^2)$. The n convex hulls can be computed with the help of the aforementioned simultaneous angular sorting from [7] followed by n calls of Graham-Sklansky linear-time procedure for the convex hull of sorted points [30]. So [2,3] can be improved to $O(n^2)$-time.

It should be noted that in [9] R8 was left as a simple exercise to be solved by a straightforward extension of the method proposed here for T9 in O(nlogn) time. But I failed to solve it immediately. It seems to me that fixed-angle wedges (see [9]) may help to solve the minmax problem with additive weights, as for multiplicative weigths one has to minimize the angle of a containing wedge which is not so simple. But, perhaps, I am wrong.

Problem R9 (constrained R8). $\max_i w_i d_E(p_i,L) \rightarrow \min_L$, where L is constrained to pass through a fixed point, say, through the origin.

After "scaling" the pointset $(x_i,y_i) \rightarrow (w_i x_i, w_i y_i)$ we obtain the equivalent unweighted constrained minmax line problem which is solved in [20] in O(nlogn) time. Within this time we can solve still more constrained version [20].

Problem R10. Find a ray R from the origin minimizing $\max_i w_i d_E(p_i,R)$.

As one can readily observe, problems R4-R10 may also find applications in location of railroads, irrigation ditches, conveyor belts etc. This role is also well-known [26,34] and in operations research these problems have another name: "location of linear facilities in the plane".

In this context a natural extension of R4 arises.

Problem R11. Minimize the sum of the cost function from R4 and the largest distance between projections of p_i onto L (the width of P in L-direction).

In [2, 3] it is solved in $O(n^2 logn)$ time, O(n) space. But we note that the characterization of L_{R11} from [2, 3] makes the method from [18] also applicable to R11 giving an $O(n^2)$-time and space algorithm for it.

The obvious generalization of all these R-problems is

Problem group R12. Locate k linear facilities optimally.

It turns out that [25] such k-line problems are NP-hard in strong sence (see [14] on NP-hardness) for any isotone cost function based on any reasonable metric if we allow k be a part of the input to the problem. The reason lies in strong NP-completeness of the following problem.

Problem R13 [25]. For any n and any k decide whether n given points can be covered by k straight lines.

However there remains an

Open problem group 02. For any n and fixed k locate k lines optimally w.r.t. n given points in the plane.

What about k=2?

Now let us turn back to regression viewpoint once more and consider the minmax analogue of R6.

Problem R14. $\max_i w_i |y_i - ax_i - b| \to \min_{a,b}$, i.e.

minimize maximal (weighted) vertical deviation of P from a line.

For the unweighted case [31] points out that M.Shamos in his thesis solved it in O(nlogn) time via the convex hull of P just as for R8*.

Problem R15 (dynamizing R14). While inserting and deleting points of P maintain a minimal line for R14.

[31] gives a method based on dynamical maintenance of convex hulls from [29] and two monotonicity properties of convex hulls resulting in O(nlogn) preprocessing time, $O(\log^2 n)$ update time and $O(\log^2 n)$ time for finding current L_{R14}.

3. Projection or shadow problems. Scene: A set of arbitrary computationally simple objects. The "simplicity" means O(1) storage description, O(1) time intersection or tangent finding etc. in various relevant combinations.

In [12] a rather general method for maintaining a parallel view of a scene when walking around it is given. As a corollary it can be applied to the following problems.

Problems P1-P5. For a given scene find a line (or a direction) such that

- the projections (shadows) of the objects onto it do not overlap
- the total projection of the scene is connected;
- the total projection of the scene is disconnected;

- it minimizes the maximal hole in the projection;
- it maximizes the minimal hole in the projection;
- etc.

Problem P6(transversal or stabbing line). Find a direction in which all projections have common point, or, equivalently, find a line piercing all objects of the scene.

(For the completeness, the diameter [30] and the minimum width strip (R8*) problems also fall into this cathegory but have O(nlogn)-time algorithms.)

For P1-P6 [12] gives a tradeoff class of $O(n^3/f(n))$-time, $O(nf(n))$-space algorithms where $1 \leq f(n) \leq n/\log n$, for a scene with nonintersecting but possibly touching simple objects. But as we will see below that for some problems faster specialized algorithms exist.

The following two problems bear relation to finding the best linear combination of hashing functions [8].

Problem P7. Consider a pointset scene P and its projection $P^a = \{p_i^a\}$ onto a line with slope angle a. The span of P^a is $\max_{i,j} |p_i^a - p_j^a|$. The resolution of P^a is $\min_{i,j} |p_i^a - p_j^a|$. The length of P^a is span/resolution. Find a slope a minimizing the length of P^a.

It is related to the next one. Given a fixed set of keys P and two functions h1, h2 mapping P to Z^+, find constants c1, c2, c3 such that the hash function h(x)=c1h1(x)+c2h2(x)+c3 produces a minimum table size perfect hashing of P.

In [8] an $O(n^2\log n)$-time, $O(n^2)$-space algorithm is given for P7. Perhaps, P7 may be amenable to the general method of [12].

Problem P8. The span of P^a is divided into a fixed number b of equal length buckets. Optimize some cost function of the distribution of P^a into the buckets, i.e. minimize maximal bucket contents.

For the latter minmax problem [8] gives an $O(n^2 b\log nb)$-time algorithm. For more general cost functions the complexity estimate must include costs of the cost function recomputations during the search.

4. Transversal and separating lines. Among the problems of the previous section the problem P6 gained more attention for different reasons, one of which is that it is a classic combinatorial problem with a number of Helly-type theorems for various classes of objects (e.g. [15], [17]) leading to $O(n^c)$-time algorithms with $c \geq 3$. Another reason is that it is related to the problem R8: instead of minimizing one can ask a decision problem.

Problem T1. Is there a line L such that $\max_i w_i d_E(p_i, L) \leq D$?

Consider n discs with radii D/w_i centered at p_i. Then the latter problem is obviously stated as the stabbing one.

Problem T2. Is there a straight line intersecting all these discs?

This problem can be solved in time O(nlogn) [4,5,6,9]. What is more, this time is sufficient to find <u>all</u> stabbing lines.

Consider now other special cases of P6.

Problem T3. Is there a line intersecting all given vertical line segments?

In statistics sometimes one needs to detect linear dependence when instead of actual values of a function only confidence limits ("bars") of these values are given. Another application in computer graphics is related to checking whether a given set of pixels represents a straight line.

In [28] an O(nlogn)-time algorithm was given for T3. However in [9] it is pointed out that this problem is a special case of

Problem T4. For two pointsets find a line separating them.

(In T3 you only need to separate bottom segment ends from top ones.)

The problem T4 is also well studied and is known to be reducible to linear programming. When the dimension is fixed linear programs can be solved in linear time [23]. So T3 is solved in O(n) time via T4.

Problem T5. Find a line stabbing all given rectangles in the plane.

As noted in [9] T5 is basically the same as T3: here the stabbing line separates either SE corners from NW corners or SW from NE or both, so one needs only to solve two problems T4. This result generalizes to any fixed-dimensional hyperrectangles (stabbed by a hyperplane) [9].

There is still other generalization of T3 in this direction. An object (in the plane) is said to be k-oriented if it is constructed from segments parallel only to k given directions.

Problem T6. Find a stabbing line for a set of k-oriented objects.

Here we also have an O(n)-time algorithm (more precisely, O(kn)) [4, 9]. You can go in d dimensions with T6 too.

In [10] you can find some other problems on separability, "ham sandwitch cut" etc.

Problem T7. Find a stabbing line for a set of n arbitrary line segments in the plane.

In [11] an O(nlogn)-time algorithm is given to find <u>all</u> stabbing lines. This paper may serve as an excellent introduction into Computational Geometry: 6 authors ingenuously combine 6 main algorithmic paradigms used in computational geometry: geometric transforms, divide-and-conquer, plane sweep, exploiting the convexity, query setting and Dynamizing.

Problem T8 (stabbing query). Preprocess a set of n line segments in order to check quickly whether a query line is stabbing.

After O(nlogn)-time preprocessing segment-stabbing line queries can be answered in time O(logn) [11]. As for dynamizing, O(n loglogn) space, O(n) update time and O(logn) query answer time are possible by standard methods [11] .

The next problem generalizes both T2 and T7.

Problem T9. Find transversal lines for a set of homothets of a simple object in the plane.

In [9] it is solved as follows. As each homothet is uniquely defined by three parameters: shifts of prototype object in x and y directions and scaling,- it can be represented by a point in E^3. Then it turns out that to a planar straight line in E^3 there corresponds a wedge of certain kind such that a line will be a stabbing for these homothets if and only if this wedge contains all points - images of homothets. All this leads to an O(nlogn)-time algorithm working in E^3 [9] .

The approach from [11] can be extended to still more general case. In [5] the following problem was considered.

Problem T10. Find all stabbing lines for a set of n computationally simple figures in the plane.

Here we have an O(n logn f(n))-time algorithm where f(n) is an extremely slowly growing fuction (such that O(n f(n)) is an upper bound for a length of a Davenport-Schinzel sequence [32]). When the boundaries of any pair of objects intersect no more than 2 times then the f(n)-term can be dropped.

Problem T11. Find all stabbing lines for a set of planar figures whose boundaries consist of computationally simple pieces where the total number of pieces is n.

The result of [4] applies to the case of nonintersecting figures with O(nlogn) algorithm. However just as in [5] you can insert the Davenport-Schinzel term f(n) to allow intersections.

As for the optimality of these stabbing algorithms, in [4] a lower bound of Omega(nlogn) is proven for a problem of finding all stabbing lines even for a set of vertical line segments, cf. T3 (the proof is by reduction from the Convex Hull problem via a geometric transform). This lower bound is not applicable to the problem of finding any stabbing line. So one may pose problems such as

Open problem O3. Is there an O(n)-time algorithm for detecting a stabbing line for a set of n planar discs?

Finally, one may ask

Problem T12. For any n and any k decide whether n given points can be stabbed by k lines.

Unfortunally this problem is NP-complete for any kind of objects, for we can replace the points in the NP-completeness proof for R13 in [25] by sufficiently small segments, circles, etc. Hence the NP-hardness of

Problem T13. For a set of objects in the plane find the minimal number of straight lines stabbing all of them.

Just as in sect.2, there still remains an

Open problem group 04. For any n and fixed k decide whether n given objects in the plane can be stabbed by k lines.(What if k=2?)

References

[1] Дорошко Н.Н.: Решение двух задач о магистралях на плоскости. Деп. ВИНИТИ I4.06.I984. Регистр. № 396IДеп. I5 с.

[2] Дорошко Н.Н., Корнеенко Н.М., Метельский Н.Н.: Оптимизация расположения прямой относительно системы точек на плоскости. Минск. Ин-т математики АН БССР. I984. Препринт № 23. I9 с.

[3] Дорошко Н.Н., Корнеенко Н.М., Метельский Н.Н.: Задачи о магистралях на плоскости. Докл.АН БССР. 29 (I985) 872-873.

[4] Корнеенко Н.М.: Поиск трансверсальных прямых для множеств выпуклых фигур на плоскости. Изв.АН БССР. Сер.физ.-мат.наук. (I986) 5, 6-I0.

[5] Atallah M.,Bajaj C.: Efficient algorithms for common transversals. Inform.Process.Lett. 25 (1987) 87-91.

[6] Bajaj C.,Li M.: On the duality of intersection and closest point problems. Proc. 21st Ann. Allerton Conf. Commun., Control, Comput. (1983) 459-461.

[7] Chazelle B., Guibas L., Lee D.T.: The power of geometric duality. Proc. 24th Ann. IEEE Symp. Found. Computer Sci. (1983) 217-225.

[8] Comer D., O'Donnell M.: Geometric problems with application to hashing. SIAM J. Comput. 11 (1985) 217-226.

[9] Edelsbrunner H.: Finding transversals for sets of simple geometric figures. Theoret. Computer Sci. 35 (1985) 55-59.

[10] Edelsbrunner H.: Algorithms in combinatorial geometry. Springer-Verlag (1987) 423+xv p.

[11] Edelsbrunner H., Maurer H., Preparata F., Rosenberg A., Welzl E., Wood D.: Stabbing line segments. BIT 22 (1982) 274-281.

[12] Edelsbrunner H., Overmars M., Wood D.: Graphics in Flatland: a case study. Adv. Comput. Res. 1 (1983) 35-60.

[13] Erdős P., Lovasz L., Simmons A., Strauss E.: Dissection graphs of planar point sets. A survey of Comb. Theory. Eds Srivastava J. et al. (1973) 139-149.

[14] Garey M.G., Johnson D.S.: Computers and intractability. Freeman San-Francisco (1979) 410 p. (see also numerous translations)

[15] Hadwiger H., Debrunner D.: Combinatorial geometry in the plane. (Various editions in many languages starting from 1959.)

[16] Houle M., Toussaint G.T.: Computing the width of a set. Proc. ACM Symp. Comput. Geom. (1985) 1-7.

[17] Katchalski M., Lewis T.: Cutting rectangles in the plane. Discrete Math. 42 (1982) 67-71.

[18] Lee D.T., Ching Y.T.: The power of geometric duality revisited. Inform. Process. Lett. 21 (1985) 117-122.

[19] Lee D.T., Preparata F.P.: Computational geometry - a survey. IEEE Trans. Comput. 33 (1984) 1072-1101.

[20] Lee D.T., Wu Y.F.: Geometric complexity of some location problems. Algorithmica 1 (1986) 193-211.

[21] Lovasz L.: On the number of halving lines. Ann. Univ. Sci.Budapest. Sekt. Math. 14 (1971) 107-108.

[22] Mandel J.: Fitting straight lines when both variables are subject to errors. J. Quality Technol. 16 (1984) 1-14.

[23] Megiddo N.: Linear-time algorithms for linear programming in R^3 and related problems. SIAM J. Comput. 12 (1983) 759-776.

[24] Megiddo N., Tamir A.: On the complexity of locating linear facilities in the plane. Oper. Res. Lett. 1 (1982) 194-197.

[25] Megiddo N., Tamir A.: Finding least-distance lines. SIAM J. Algebraic and Discrete Meth. 4 (1983) 207-211.

[26] Morris J.G., Norback J.P.: A simple approach to linear facilities location. Transp. Sci. 14 (1980) 1-8.

[27] Morris J.G., Norback J.P.: Linear facility location - solving extensions of basic problem. Eur.J. Oper. Res. 12 (1983) 90-94.

[28] O'Rourke J.: An on-line algorithm for fitting straight lines between data ranges. Commun. ACM. 24 (1981) 574-578.

[29] Overmars M.H., van Leeuwen J.: Dynamically maintaining configurations in the plane. J.Comput.Syst.Sci.23 (1981) 166-204 .

[30] Preparata F.P., Shamos M.I.: Computational geometry: an introduction. Springer-Verlag. (1985) 390+xii p.

[31] Rey C., Ward P.: On determining the on-line minmax linear fit to a discrete pointset in the plane. Inform. Process. Lett. 24 (1987) 97-102.

[32] Sharir M., Livne R.: On minima of functions, intersection patterns of curves and Davenport-Schinzel sequences. Proc. 26th Ann. IEEE Symp. Found. Computer Sci. (1985) 312-320.

[33] Szemeredi E., Trotter W.: Extremal problems in discrete geometry. Combinatorica. 3 (1983) 381-392.

[34] Wesolowsky G.O.: Location of the median line for weighted points. Environ. Planning A7 (1975) 163-170.

SPECIAL DISSECTION OF POLYGONAL REGIONS IN LSI MASK PATTERN DESIGN

L.V.Nosov[1], N.M.Korneenko[2]

In programming for pattern generator (in particular for electron-beam) a number of problems concerned with a special dissection of polygonal regions into elementary figures arises. This paper deals with one of the most typical of such problems. The fast algorithm is presented. Its time complexity is investigated. The proposed method may be applied to other special dissections.

1. Introduction

Given a plane with a digitized coordinate system and a region A closed with $N(A)$ straight line segments. A polygonal region will be named also as region or figure. $Re(A)$ is the bounding rectangle with sides parallel to coordinate axes. Consider on X and Y axes respectively two increasing sequences of numbers $X=\{x_j \mid j\in I\}$, $Y=\{y_i \mid i\in I\}$. The lines $y=y_i$, $x=x_j$, $x_j\in X$, $y_i\in Y$, generate the orthogonal net. Denote by A_{y_i} (A^{x_j}) an intersection of A with a stripe, restricted by lines $y=y_i$ and $y=y_{i+1}$ ($x=x_j$ and $x=x_{j+1}$). Let $A^{x_j}_{y_i}=(A_{y_i})^{x_j}=(A^{x_j})_{y_i}$.

Denote by $N_Y(A)(N_X(A))$ the number of intersections of non-horizontal (non-vertical) boundary segments of figure A with the lines $y=y_i$, $y_i\in Y$ ($x=x_j$, $x_j\in X$). Let $N_{Y,X}(A)$ is a total number of intersections of lines $y=y_i$ with lines $x=x_j$, $x_j\in X$, $y_i\in Y$, lied in A. If it will not lead to misunderstandings then N, N_X, N_Y, $N_{Y,X}$ will be used instead of $N(A)$, $N_X(A)$, $N_Y(A)$, $N_{Y,X}(A)$, respectively. The problem of special dissection (SD problem) is formulated as follows.

SD problem. Given a polygonal region A and the following sequences of numbers.

$$X=\{x_j \mid x_j=\delta_x\cdot j+\rho_x,\ 0\le\rho_x<\delta_x,\ j\in I\},\tag{1}$$

$$Y=\{y_i \mid y_i=\delta_y\cdot i+\rho_y,\ 0\le\rho_y<\delta_y,\ i\in I\},\tag{2}$$

where I is the set of integers. It is necessary

1) to dissect figures $A^{x_j}_{y_i}$ into trapezoids;

2) to sort trapezoids lexicographically on increasing y_i, x_j (each tra-

---

[1] СССР, 220027, Минск, Ленинский проспект, 65/I, Белорусский политехнический институт, кафедра САПР.

[2] СССР, 220604, Минск, 72, ул. Сурганова II, Институт математики АН БССР

pezoid is the dissection element of the figure $A_{y_i}^{x_j}$ and y_i, x_j, are the indexes of such figure).

The purpose of this paper is to solve the SD problem in $O(N_Y + N_X + N_{Y,X} + N \log N)$ time and to prove that this time complexity is invariant with regard to location of the figure and the orthogonal net.

The plane-sweep algorithm dissecting a figure into trapezoids in $O(N \log N)$ time is presented in [3] . But a direct application of a plane-sweep method to solve the SD problem doesn't allow to lead to the desired time complexity. The proposed solution of SD problem is based on a combination of the plane-sweep method and its modification.

2. Plane-sweep method and its modification

Let $R = \{r_k\}_{k=1}^{N}$ is a set of pairwise non-intersecting non-horizontal straight line segments. We shall restrict our consideration of a plane-sweep method by sweeping of R because the sweeping of non-horizontal boundary segments of a polygonal figure differs from it only in insignificant details [3]. Sketch in few words the sweeping of R presented in [4]. Introduce on the set R the following relation. Two segments r_i and r_j are comparable if there exists a horizontal line that intersects both of them. We shall say that r_i is more left than r_j (and r_j is more right than r_i) if the abscissa of intersection of r_i with a horizontal line is less than the abscissa of intersection of r_j with this line. Note that in this way the order relation is introduced on the set of segments intersecting with a horizontal line. Let $P = \{ p(j) \mid j = 1, 2, \ldots, 2N \}$ is the sorted lexicographically on increasing coordinates y, x list of endpoints of segments from R ; T is a structure to maintain the ordering (at first T is empty); procedure INSERT(r,T) inserts the segment r in T and procedure DELETE(r,T) deletes r according to introduced order relation. The sweeping of R is as follows.

```
      begin
   for j←1 until 2N do begin
      if p(j) is the lowest end of segment r
        then INSERT(r,T),
        else DELETE(r,T);
      end              end
```

T usually is a balanced search tree [1]. Thus the total time of the sweeping of R is equal to $O(N \log N)$.

In modification of the plane-sweep method one chooses the structure more complicated than balanced search tree. This allows to process in linear time the segments with both endpoints on the boundary of $Re(R)$. $Re(R)$ is the bounding rectangle of R with sides parallel to coordinate axes.

Theorem 2.1. If the endpoints of the segments of R are preliminary sorted lexicographically on increasing coordinates y,x then $O(N'+N''\log N''+N''\log N')$ time is sufficient to sweep R. N' is the number of the segments with both endpoints on the boundary of $Re(R)$, N'' is the number of remaining segments.

The proof yields from the following observation. In sweeping of the set of segments with both endpoints on the boundary of $Re(R)$ one inserts (deletes) segments only in the beginning or in the end of T. Thus T may be a list and the sweeping time is $O(N')$.

In the common case T is the table $\bar{R}=\{\bar{r}_0,\bar{r}_1,\dots,\bar{r}_{n+1}\}$ and the set of balanced search trees $\{T_{i,i+1}\}_{i=0}^{n}$ associated with it. The table $\bar{R}$ is for presentation of segments with both endpoints on the boundary of $Re(R)$. Each tree $T_{i,i+1}$ is for presentation of segments which are more right than $\bar{r}_i$ and more left than $\bar{r}_{i+1}$ with at least one endpoint in $Re(R)$ (fig. 1). Without lost of generality we assume that in $\bar{R}$ there is always a leftmost segment $\bar{r}_0$ and a rightmost segment $\bar{r}_{n+1}$.

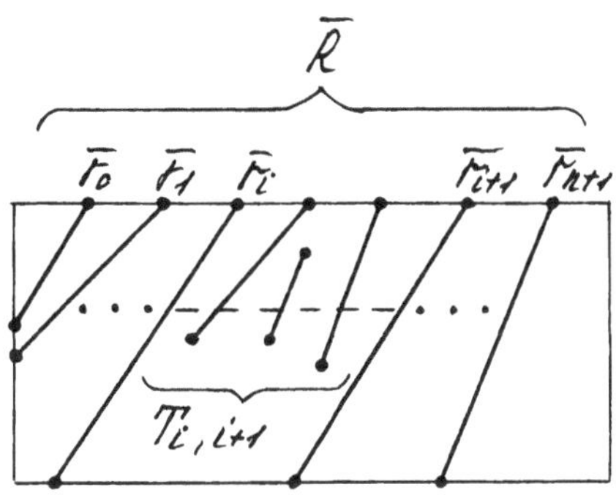

fig. 1. To the proof of the theorem 2.1.

Then the processing of segment r with at least one endpoint in $Re(R)$ consists firstly in searching in $\bar{R}$ a pair of elements $\bar{r}_i$ and $\bar{r}_{i+1}$ (r is more right than $\bar{r}_i$ and more left than $\bar{r}_{i+1}$) by the "divide and conquer" method and secondly in inserting (deleting) r in the appropriate tree $T_{i,i+1}$. It takes in the worst case $O(\log N'+\log N'')$ time. Thus the total sweeping time is $O(N'+N''\log N''+N''\log N')$.

3. Solving of the SD problem

The purpose of this paragraph is to prove the following theorem.

Theorem 3.1.

1) There exists an algorithm that solves the SD problem in $O(N_Y+N_X+N_{Y,X}+N\log N)$ time;

2) $O(N_Y+N_X+N_{Y,X}+N\log N)$ is invariant relatively to a location of the fi-

gure and the orthogonal net.

Consider the auxiliary problem of dissection a figure into horizontal stripes (HSD problem).

HSD problem. Given figure A and the sequence of numbers (2). It is necessary

1) to construct figures A_{y_i}, $y_i \in Y$, in the form of a list of non-horizontal boundary segments;

2) to sort the output list lexicographically on increasing y_i, y, x, where y_i is the index of a stripe that contains the segment.

Lemma 3.1. There exists a plane-sweep algorithm solving the HSD problem in $O(N_Y+t_s+t_{sw})$ time (assuming that $\lceil\ \rceil$ is an elementary operation), where $t_s = N\log N$ is the time of the preprocessing sorting of the endpoints of boundary segments of the figure; t_{sw} is the sweeping time of non-horizontal boundary segments of A (in an ordinary sweeping $t_{sw} = N\log N$).

This algorithm one can find in [5].

Execute the following sequence of actions.

1) Solve for the input figure the HSD problem. It takes $O(N_Y + N\log N)$ time by lemma 3.1;

2) Form the list of the endpoints of non-vertical boundary segments of figures A_{y_i}, $y_i \in Y$. For this add to the input list the endpoints of horizontal boundary segments of figures A_{y_i}, $y_i \in Y$, and delete the endpoints of vertical segments. It takes only $O(N_Y+N)$ time since the input list of the endpoints of non-horizontal boundary segments of figures A_{y_i}, $y_i \in Y$, is sorted lexicographically on increasing y_i, y, x.

3) Exchange X and Y axes (then a sequence Y becomes a sequence X, figures A_{y_i} become figures A^{x_j}, horizontal segments become vertical) and solve the HSD problem for the set of figures A^{x_j}, $x_j \in X$. Note that the preprocessing sorting takes $O(N_X + N\log N)$ time because the lists of the endpoints of vertical boundary segments of figures A^{x_j} lied on the lines $x = x_j$, $x_j \in X$, are already sorted. Use the modification of the plane-sweep method in solving the HSD problem. According to theorem 2.1 and lemma 3.1 $O(N_Y+N_X+N_{Y,X}+N\log N)$ time is sufficient for this;

4) Apply the mentioned above plane-sweep algorithm from [3] to dissect the figures $A^{x_j}_{y_i}$, $x_j \in X$, $y_i \in Y$, into trapezoids. If one uses the modification of a plane-sweep method then the total time of dissection equals to $O(N_Y+N_X+N_{Y,X}+N\log N)$.

The first part of theorem 3.1 is proved. The detailed proof of the second part is presented in [2]. So we formulate the main lemmas from [2] without a proof. Below X and Y are the sequences (1), (2), respectively.

Lemma 3.2. Let A' denotes the location of figure A after a parallel shift. Then the following unequalities are true: $N_X(A) \le N_X(A') + N$; $N_Y(A) \le N_Y(A') + N$; $N_{Y,X}(A) \le N_{Y,X}(A') + 1/2\,(N_Y(A') + N_X(A') + N)$

Lemma 3.3. For any two figures A and A' of the same area the following equality takes place $N_{Y,X}(A) = O(N_{Y,X}(A') + N_Y(A') + N_X(A') + N_Y(A) + N_X(A) + N(A) + N(A'))$.

Lemma 3.4. Let A' denotes the location of figure A after any rotation. Then the equality $N_Y(A) + N_X(A) = O(N_Y(A') + N_X(A') + N)$ is true.

Literature

[1] Адельсон-Бельский Г.М., Ландис Е.М.: Один алгоритм организации информации. ДАН СССР. I46 (I962), 263-266.

[2] Носов Л.В.: Алгоритмическая сложность задач построения специализированных покрытий плоских многоугольных областей. Деп. в ВИНИТИ 24.06.87, № 46I8-B87.

[3] Носов Л.В., Носова Е.Г., Рабинович Е.Б., Фейнберг В.З.: Оптимальные алгоритмы взаиморасположения и покрытия плоских многоугольных областей. Изв. АН БССР. 4 (I985), 28-36.

[4] Shamos M.I., Hoey D.: Geometrical intersection problems. 17 Annual Symp. Found. Comput. Sci. (1976), 208-215.

[5] Фейнберг В.З.: Геометрические задачи машинной графики больших интегральных схем. Сов. радио. Москва, I987.

SEPARABILITY OF SIMPLE POLYGONS

Günter Werner [1)]

Abstract

Separability is a special case of motion planning problems. In the paper two types of separability are examined: For a natural number k the term "k-separability" is used, if two simple polygons in the Euclidean plane are separated by, at most, k translations only. The special case of 1-separability has already been intensively examined. We give complexity results for the general case and the m-separability, which is a minimal length separability with translations. Algorithms for deciding the different types of separability and determining the separations are proposed.

1. Introduction

Motion problems are manifold /6,7,9,10,12,13,15,16,17/ due to the variety of areas in which they may occur; among these areas we find e.g. robotics, computer graphics, computer vision, computer aided design etc. A famous survey concerning problems, methods for solution, complexity results is given in /20/.

One class of motion problems is that of separability problems. To state this problem formally, we must introduce some terminology. Consider two simple polygons P,Q in the Euclidean plane, with non-intersecting interiors. P and Q are linear separable, if a line l between P and Q exists, i.e. P and Q are on opposite sides of this line l. Two simple polygons are linear separable if and only if their convex hulls are linear separable. Furthermore, two convex polygons are linear separable if and only if they do not intersect. The computation time of the convex hull of a simple polygon /8/ and the polygon-intersection-test are linear in the number of vertices of the polygons. Consequently, the linear separability of two simple polygons is decidable in linear time.

A translation of the polygon P is specified by the translation direction and its distance. A separating motion of P is a set of translations of P without colliding with the polygon Q, such that finally P and Q are linear separable. P is said to collide with Q, if at any translation of the separating motion the interiors of P and Q

---

1) Friedrich-Schiller-Universität Jena, DDR-6900 Jena,
Sektion Mathematik, Universitätshochhaus 17.OG

intersect, otherwise P and Q are said to be separable by a sequence of translations. P and Q interlock, if they are not separable by a sequence of translations.

Separability problems are classified into problems of detection and determination. The first problem is to decide, whether the objects are separable in the defined sense, while the latter problem to the determine the separating motion.

The paper is organized as follows. In Section 2 we summarize some results concerning separability with a single translation. In Section 3 and 4 we discuss the computation of a sequence of separating translations with additional conditions. In Section 3 this condition consists in the minimality of the sums of all translation distances. This separability is called the m-separability.

In the next Section we examine the separability by a minimal number of translations. For a natural number k, the term "k-separability" denotes the separability with, at most, k translations (see fig. 1).

2. The 1-separability

In the case of the 1-separability the detection and the determination problems have to be modified in the following sense. Since one translation at most is allowed for separating the two simple polygons, the detection problem is split into two tasks:

A1: Given a direction A. Is it possible to separate the polygons by a translation in direction A?

A2: Is it possible to separate the polygons by a translation in an arbitrary direction?

In general there is more than only one translation with an infinite distance, which separates two 1-separable polygons. The set of all directions of these translations, 1-separating two simple polygons P and Q, is denoted by W(P,Q). The so-called movability wedge W(P,Q) /10/ is the union of the relative movability wedges W(Q) and W(P), where W(Q) is the set of all directions of translations, 1-separating P and Q, P is fixed. Clearly, W(Q) can be obtained from W(P) by a 180° rotation of the wedge. Of course, if W(P,Q) is empty, then P and Q interlock.

We assume that both polygons have in order the same number of vertices, say n. Table 1 gives a survey about some complexity results for solving the detection problems A1 and A2 and the determination problem for simple polygons and other more restricted classes of polygons /15,16,17/. The techniques used range from standard tools of computation al geometry combined with binary search for convex polygons /5/, the computation of the visibility hull for simple polygons /4/ to partitioning techniques for computing the movability wedges of simple polygons /10/.

Polygon class	Detection		Determination
	A1	A2	W(P,Q)
constant size	O(1)	O(1)	O(1)
convex	O(logn)	O(1)	O(logn)
monotone	O(n)	O(1)	O(n)
star-shaped	O(n)	O(1)	O(n)
simple	O(n)	O(n)	O(nlogn)

Table 1: Complexity results of detection and determination problems for 1-separability

3. The m-separability

A standard technique in motion planning is the use of the configuration space, introduced by Lozano-Perez and Wesley /6/. The configuration space of two sets, A and B, is A-B={a-b:a∈A, b∈B}. Assume that the origin is the startpoint of a movement and is also the reference point of an object P. P can be moved by a set of translations without colliding with an obstacle Q from a start- to a goal position if, and only if, both the start and the reference point of the goal position are in the same hole of the configuration space Q-P or if they are in the unbounded region of Q-P.

The configuration space of two simple polygons may have $\Omega(n^2)$ vertices. A lower bound for the time complexity for computing the configuration space has been proposed by Lozano-Perez and Wesley /6/. If the polygons P and Q are convex, then a linear time algorithm for the computation of the configuration space has been developed by Lozano-Perez /7/. The idea he uses also appears in Schwartz /11/.

We may now state the following Lemma:

Lemma 1: Two simple polygons are m-separable if and only if the origin is not included in the configuration space of the polygons.

Once the configuration space has been computed, the rest is done by a point inclusion test with the border of the unbounded region of the configuration space. Thus we have:

Theorem 1: Given two simple polygons P and Q with O(n) vertices. The m-separability of P and Q is decidable in $O(n^2)$ time.

Now we assume, that P and Q are m-separable and, that P and Q are not linear separable. The determination of the m-separating sequence of translations we obtain as follows. When the configuration space P-Q is computed, then the shortest nonintersecting path of the origin to a pseudo edge of the convex hull is exactly the m-separating sequence. A pseudo edge is an edge, which is not an edge in the original polygon. The sequence of translations so computed moves Q away from P, i.e. P is fixed. With the sequence of inverse translations P will be moved away from Q.

There are different techniques for computing this path. For a survey see /20/. Another effective method has been developed by Chazelle /2/. As a preprocessing step he uses a triangulation of the convex hull of the configuration space. This is possible in O(vloglogv) time /14/, where v is the number of vertices of the configuration space. Using triangulation the m-separating sequence is computed in O(v) time. The convex hull of the configuration space is also computed in O(v) time /8/. Observe that only the convex hull of the border of the unbounded region has to be computed. We obtain an O(vloglogv) algorithm for the m-separating determination problem, where $v=O(n^2)$.

Theorem 2: The m-separating path for two simple polygons with O(n) vertices is determined in $O(n^2 \log\log n)$ time.

4. The k-separability

Two detection problems are to be examined:

A1: Is there a natural number k, such that the polygons P and Q are k-separable?

A2: For a given number k, are the polygons P and Q k-separable?

Obviously, the problem A1 is equivalent to the detection problem of the m-separability, thus A1 is soluble in $O(n^2)$ time.

The determination problem for the k-separability can also be devided into two problems:

B1: Let the polygons P and Q be m-separable. Determine the smallest number k, such that P and Q are k-separable.

B2: Let the polygons P and Q be k-separable, and k be minimal. Determine a sequence of translations, which k-separates P and Q.

The k-separability is related to the weak visibility /18,19/. In the configuration space the visibility polygon from the origin /4/ describes the set of all positions, to which the origin is movable by one translation. Let e be a pseudo edge of the visibility polygon, that splits the configuration space into two parts A and B. Assume that A contains the origin. Then the weak visible part of B from the edge e is the set of all positions, which can be reached from the origin by two translations. Continuing this process we get after k iterations, the set of all those positions, which can be reached from the origin by k translations. This idea is successful for solving the detection and the determination problem of the k-separability.

Let P and Q be not linear separable and CS=P-Q be the configuration space. Then the convex hull of CS contains the origin. Let us further assume, that P and Q are m-separable, i.e. the origin is not in a hole of CS.

Clearly, the detection problem, A2, has to be answered positively if and only if after, at most, k iterations which are described above, a point of a pseudo edge of the convex hull of the configuration space is weakly visible. Otherwise the minimal number k, which gives a positive answer to the problem A2, is the minimal number of such iterations. Thus a method for the solution of the problem B1 is given. Now let k be the minimal number of translations which k-separate P and Q. A k-separating sequence of translations for a given P and Q is not unique. The result is the following sequence: Starting with the origin, we select, for each iteration step, a point from the interior of the respective weak visibility polygon. The sequence of points, obtained in this way describes a sequence of translations, k-separating P and Q.

In order in examine the complexity we must observe, that the visibility polygon from a point is computed in O(v) time, where v is the number of vertices of the polygon /4/, while the weak visibility polygons are recursively computed in linear time /1,3/, once a triangulation /14/ is made. The computation of the convex hull of the configuration space is dominated by the visibility computations. Thus we may state the following Theorem:

Theorem 3: P and Q are simple polygons with O(n) vertices. The detection problem A1 is solvable in $O(n^2)$ time, while the problems A2, B1, B2 are solvable in $O(n^2 \log\log n)$ time.

References

/1/ Avis,D., Toussaint,G.T.: An optimal algorithm for determining the visibility of a polygon from an edge, IEEE Trans. on Comp. vol. C-30, 1981, pp. 910-914

/2/ Chazelle,B.: The Power of Triangulation: Applications to Problems of Visibility and Internal Distance, TR, Dep. of Comp. Sc., CMU, 1982

/3/ El Gindy,H.: Efficient algorithms for computing the weak visibility polygon from an edge, TR, School of Computer Science, McGill University, 1985

/4/ El Gindy,H., Avis,D.: A linear algorithm for computing the visibility polygon from a point, Journal of Algorithms, vol.2, 1981, pp. 186-197

/5/ Kirkpatrick,D.: Efficient computation of continuous skeletons, Proc. 20th IEEE Symp. on Foundations of Comp. Sci., 1979. pp. 18-27

/6/ Lozano-Perez,T., Wesley,M.: An algorithm for planning collision-free paths among polyhedral obstacles, Comm. ACM 22, 1979, pp. 421-427

/7/ Lozano-Perez,T.: Automatic planning of manipulator transfer movements, IEEE Trans. Syst., Man., Cybern., vol. SMC-11, 1981, pp. 681-698

/8/ McCallum, D., Avis,D.: A linear algorithm for finding the convex hull of a simple polygon, Info. Proc. Lett. 9, 1979, pp. 201-206

/9/ O'Dunlaing,C., Yap,C.K.: The Voronoi method for motion planning: I. The case of a disc, TR No. 53, Courant Institute, N.Y.U., 1983

/10/ Sack,J.-R., Toussiant,G.T.: Translating polygons in the plane, Proc. STACS'85, Saarbrücken, 1986, pp. 310-321

/11/ Schwartz,J.T.: Finding the minimum distance between two convex polygons, Inform. Proc. Lett., vol. 13, 1981, pp. 168-170

/12/ Schwartz,J.T., Sharir,M.: On the piano movers'problem: II. General techniques for computing topological properties of real algebraic manifolds, Adv. Appl. Math. 4, 1983, pp. 298-351

/13/ Spirakis,P., Yap,C.K.: Strong NP-hardness of moving many discs, TR No. 92, Courant Institute, N.Y.U., 1983

/14/ Tarjan,R., Van Wyk,C.: An O(nloglogn) time algorithm for triangulating simple polygons, preprint (submitted to SIAM J. Computing), 1986

/15/ Toussaint,G.T., El Gindy,H.: Separation of two monotone polygons in linear time, Robotica, vol. 2, 1984, pp. 215-220

/16/ Toussaint,G.T.: Movable separability of sets, in "Computational Geometry", Ed. G.T. Toussaint, North Holland, 1985, pp.335-376

/17/ Toussaint,G.T.: Shortest path solves translation separability of polygons, TR SCS-8.27, School of Computer Science, McGill University, 1985

/18/ Toussaint,G.T.: Pattern recognition and geometrical complexity, Proc. Fifth Intern. Conf. on Pattern Rec., vol 2, 1980, pp. 1324-1347.

/19/ Werner,G.: The weak visibility, Computers and Artificial Intelligence, No. 2, 1987, pp. 151-172

/20/ Whitesides,S.: Computational geometry and spatial planning, in "Computational Geometry", Ed. G.T. Toussaint, North Holland, 1985, pp. 377-428

Figures

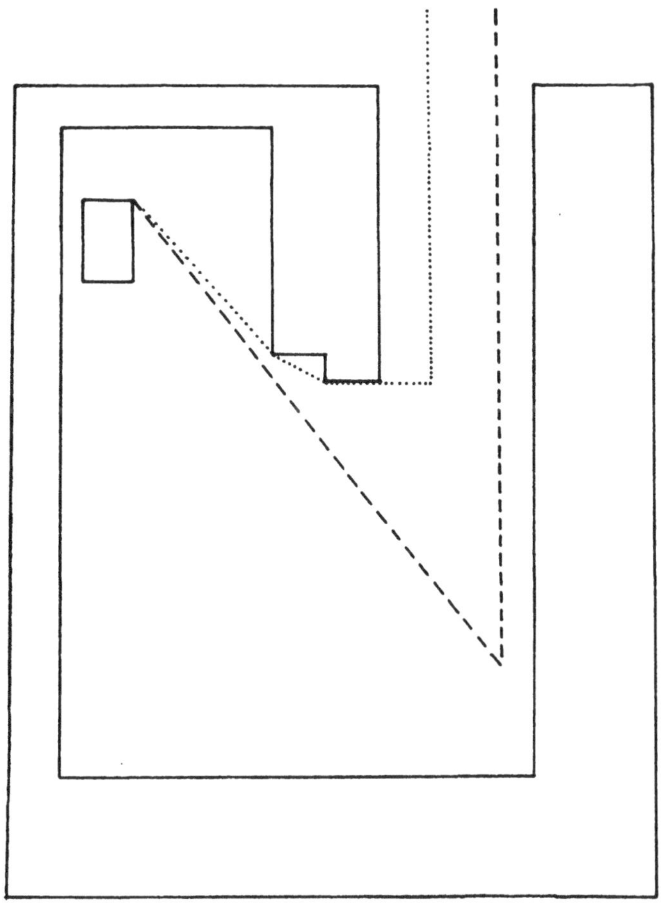

Fig. 1: The dotted line describes the m-separating sequence of translations, the dashed line is a 2-separating path.

POLARITY OF POLYHEDRA AND SEPARABILITY OF POINT SETS

Peter-Michael Schmidt [1)]

Abstract: The polarity of polyhedra is an useful tool in Computational Geometry. To apply the polarity, it is usually assumed that the origin of the coordinate system lies in the interior of the polyhedron. The problem of finding a commom interior point of a finite set of half-spaces is considered. This problem is closely related to separability of point sets and the Minkowski sum of convex polyhedra. The paper describes an optimal algorithm for constructing the Minkowski sum of two convex polygons in the plane, and develops the implementation of this algorithm.

1. Polarity of polyhedra and related problems in Computational Geometry

Let K be a subset of the d-dimensional euclidean space R^d.
The polar set K^* of K is defined by $K^* := \{x \in R^d : \langle x,y \rangle \leq 1 \text{ for all } y \in K\}$.
The polarity has many interesting and useful applications in the Combinatorial and Computational Geometry. For example, using the following equation

$$[\mathrm{conv}(K_1 \cup \ldots \cup K_n)]^* = K_1^* \cap \ldots \cap K_n^* \quad (K_1,\ldots,K_n \text{ are convex sets})$$

the Intersection Problem of Half-Spaces is transformed into the Convex-Hull Problem (If K_i is a single point, then K_i^* is a closed half-space.).
In order to apply the polarity to (convex, compact) polytopes we wish that P^* is a polytope if P so is. Therefore we restrict the polarity to the family of all polytopes containing the origin O in its interior. Under these assumptions $P^{**} = P$ holds. These are basic results in Convexity, compare [3] section 3.4. and [4]. Consequently the following computational problem arises:

> Given n half-spaces in R^d (each bounded by supporting hyperplanes of the facets of a given polytope), construct an common inner point, or establish that no such point exists.

Let T be the geometric transform that maps the
hyperplane $x_d = h_1 x_1 + \ldots + h_{d-1} x_{d-1} + h_d$ into the point $(h_1,\ldots,h_d)$
and the
point $(p_1,\ldots,p_d)$ into the hyperplane $x_d = -p_1 x_1 - \ldots - p_{d-1} x_{d-1} + p_d$.
Without loss of generality we assume that no hyperplane is vertical, that is, all given planes are not parallel to the x_d-axis. Although this assumption has only technical meaning it is very important for investigation of complexity of geometric problems (cf.[6]). Therefore the set of

1) Dr. rer. nat. Peter-Michael Schmidt
Sektion Mathematik der Friedrich-Schiller-Universität Jena
DDR - 6900 Jena Universitätshochhaus 17. OG.

all half-spaces can be partitioned into twn subsets:

the lower half-spaces

$L_i: x_d \leq h_{i,1} x_1 + \ldots + h_{i,d-1} x_{d-1} + h_{i,d} \quad (i=1\ldots k)$

and the upper half-spaces

$U_i: x_d \geq l_{i,1} x_1 + \ldots + l_{i,d-1} x_{d-1} + l_{i,d} \quad (i=1\ldots m)$

The boundering hyperplanes are mapped by T into the point sets

$A = \{(h_{i,1},\ldots,h_{i,d}): i=1\ldots k\}$ and $B = \{(l_{i,1},\ldots,l_{i,d}): i=1\ldots m\}$.

Clearly, if $p = (p_1,\ldots,p_d) \in \text{int } L_i$ then

$T(\text{bd } L_i) \in T(p)^+ := \{x \in R^d: x_d > -p_1 x_1 - \ldots - p_{d-1} x_{d-1} + p_d\}$

and analogous $p \in \text{int } U_i$ implies

$T(\text{bd } U_i) \in T(p)^- := \{x \in R^d: x_d < -p_1 x_1 - \ldots - p_{d-1} x_{d-1} + p_d\}$.

In other words: T(p) strictly separates A from B, that is, A and B are contained in different open half-spaces determined by T(p). It is a well-known result in Computational Geometry that for constructing a separating plane $O(n \cdot \log n)$ time is necessary and sufficient, where $n=k+m$.

In order to find a separating plane of the point sets A and B we construct the convex hulls of A and B and determine a pair of points (p,q) of the boundary of conv A and conv B such that the minimum distance is realized by p and q (cf. [1]).

If this distance is nonzero than the hyperplane intersecting orthogonally the line segment $\overline{pq}$ and passing through its midpoint strictly separates the point sets A and B.

This paper develops an other method for finding a separating line of two convex polygons. The corresponding algorithm can be easily imple-mented and is a obvious consequence of familiar géometric facts.

2. Minkowski sum and separability of convex polyhedra

A and B are arbitrary convex point sets of R^d. As usual,

$A + B := \{a+b: a \in A \text{ and } b \in B\}$ denotes the Minkowski sum of A and B.

The principal idea for constructing a separating plane is to determine a linear mapping $g: R^d \rightarrow R^1$ such that $g(A+(-B))=\{g(a-b): a \in A \text{ and } b \in B\}$ is a set of positive numbers.

Then $g^{-1}(\lambda)$ is the desired separating plane, where

$\max g(A) < \lambda < \min g(B)$. Thus the problèm of constructing a separating plane is reduced to the following two problems:

1. Given two convex polytopes, find its Minkowski sum.
2. A convex polytope P with $0 \notin P$ is given. Construct a linear mapping $g: R^d \rightarrow R^1$ such that g(P) is a set of positive numbers.

For that reason the next section considers the computational complexity of the first problem in the plane.

3. Constructing the Minkowski sum of two convex polygons

Let A and B been two convex polygons with k and m vertices, respectively. An algorithm and its implementation are described that solves this problem in time O(n), where n=k+m. The principal algorithmic idea is well-known and appears, for example, in [7] (p.386-387). In general A+B has exactly n vertices (All edges of A,B are not parallel.). Therefore the algorithm is optimal in the worst case.

Because $A + B = \text{conv } V$ with $V = \{a+b:\ a \in \text{vert}(A) \text{ and } b \in \text{vert}(B)\}$ we must search for the vertices of A+B in the set V. Unfortunately, the cardinality of V is in $O(n^2)$.

The sequences of vertices of the polygons A and B are stored in connected circular lists, that is, NEXT-A(a) and NEXT-B(b) denote the next vertex of a,b with respect to counterclockwise order of the vertices of A and B, respectively. $\varphi_1^A, \ldots, \varphi_k^A, \varphi_1^B, \ldots, \varphi_m^B$ denote the angles between the x_1-axis and the normal vectors of all edges of A and B. Since A and B are convex we are able to compute the following lists in linear time:

$$0 \leq \varphi_1^A < \ldots < \varphi_k^A < 2\pi \text{ and } 0 \leq \varphi_1^B < \ldots < \varphi_m^B < 2\pi.$$

These ordered lists are merged into a queue FACETS [1..n]. The algorithm successively computes the vertices of A+B and stores them in array SUM [1..n]:

```
procedure MINKOWSKI-SUM(A,B);
 choose the vertices a,b of A,B, respectively, which have maximal
 x1-coordinate, and if such a vertex belongs to a vertical edge
 choose the lower endpoint;

for i:=1 to n
  do  SUM [i] := a+b;
      if ( The oldest element belongs to a facet of A. )
         then  a := NEXT-A(a)  else  b := NEXT-B(b);
      remove the oldest element from FASETS;
 od;
delete all points SUM [i] which are not vertices of A+B
     (That means φs^A = φt^B for i=s+t.).
```

The preceding considerations are summarized in the following

Theorem: The Minkowski sum of two convex polygons A,B in the plane can be computed in optimal time O(n) where n is the sum of the numbers of vertices of A and B.

Minkowski sum of a pentagon and a triangle

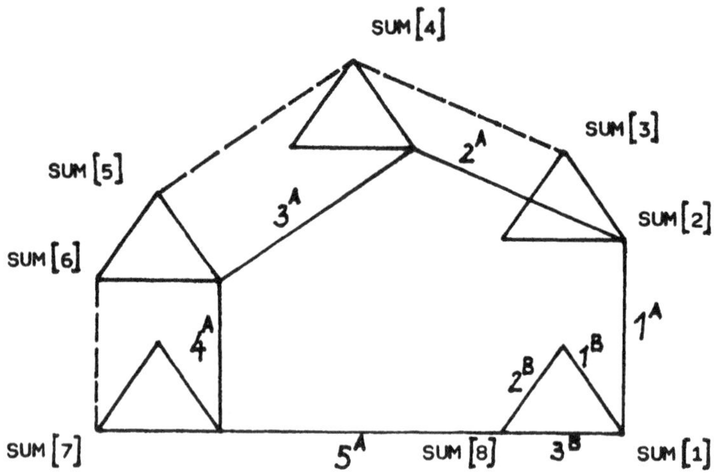

$$0 = \varphi_1^A < \varphi_1^B < \varphi_2^A < \varphi_3^A < \varphi_2^B < \varphi_4^A < \varphi_5^A = \varphi_3^B < 2\pi$$

4. References

[1] H. Edelsbrunner: Computing the Extreme Distances between Two Convex Polygons. Jour. Alg. 6 (1985), 213-224 .

[2] H. Edelsbrunner: Algorithm in Combinatorial Geometry. Springer 1987.

[3] B. Grünbaum: Convex Polytopes. London 1967.

[4] K. Leichtweiss: Konvexe Mengen. Berlin 1980.

[5] K. Mehlhorn: Data Structures and Algorithms. Vol. 3, Springer 1984.

[6] W. Nef and P.-M. Schmidt: Computing a Sweeping-Plane in Regular ('General') Position: A Numerical and a Symbolic Solution. (to appear).

[7] S. H. Whitesides: Computational Geometry and Motion Planning. Computational Geometry, Mach. Intell. Pattern Recognition 2 (1985), 377-427.

THE COMPLEXITY OF POLYGON COVER AND PARTITION PROBLEMS

Hans-Dietrich Hecker[1)]

1. Introduction

In combinatorial geometry the so-called "Art gallery problem" has caused considerable interest in the last 15 years. In 1973, VICTOR KLEE (s. [9]) posed the problem of determining the minimum number of guards sufficient to cover the interior of an n-wall art gallery room. V. CHVATAL [2] , S. FISK [3] , M. KEIL [7] , J. SACK [11]a.o. have published results, which show the connection of this problem with cover- and partition-problems in the theory of polygons. The theory has been further developed by computational geometry scientists studying partitioning algorithms. We show the NP-completeness of some minimum cover problems and give other known results on that line.

2. Definitions and notation

A simple polygon $P = (v_0, v_1, \ldots, v_n)$ is a closed plane figure whose boundary is composed of straight line edges (v_i, v_{i+1}), $i = 0,1,\ldots,n-1$, $v_0 = v_n$, and where no two nonconsecutive edges intersect. The edges of the polygon are oriented counterclockwise.

Let L denote a certain set of subsets of the polygon P. We say that L is a cover of P, if for all points $p \in P$ there exists a point $q \in L$ such that the line segment $\bar{pq}$ connecting p and q lies completely in P (p and q are visible from each other). A cover L of P is called an e-cover (p-cover), if L is a set of (single-element sets of) vertices (resp. points) of P. Similarly, a d-cover contains diagonals of the polygon only, a t-cover contains triangles with vertices v_i of P. A d^*-cover only contains diagonals, edges or vertices and a t^*-cover only contains triangles with vertices v_i of P, diagonals, edges or vertices of P.

In natural way a cover L gives a decomposition of the polygon into smaller subsets:

$$P = \bigcup_{M \in L} \{ q \in P \mid \exists_{p \in M} \ pq \subseteq P \} .$$

1) Pädagogische Hochschule Karl Friedrich Wilhelm Wander Dresden, Wigardstraße 17, Dresden DDR- 8060

A decomposition of P is called a partitioning if P is decomposed into nonoverlapping pieces. A decomposition of P is called regular, if the sets of the decomposition are polygons with vertices of the polygon P only (that means so-called Steiner-points are not allowed).
Let m(L) (resp. m(D)) denote the number of pieces (sets) in a cover L (decomposition D). We are interested in the minimum of the numbers m(L) for covers and in the minimum of the numbers m(D) for decompositions of a polygon P. Of special interest are decompositions into star-shaped pieces.

3. Complexity results

In 1985, M. KEIL has shown the following theorem for simple polygons with n vertices and N reflex vertices.

Theorem 1 [7]: There exists an $O(n^5N^2\log n)$-time algorithm for regular decomposing a simple polygon into the minimum number of star-shaped components.

In contrast, A. AGGARWAL has shown that the following theorem holds.

Theorem 2 [1]: The problem of decomposing a simple polygon into a minimum number of star-shaped pieces is NP-complete.

For $r \in \{d, t, d^*, t^*\}$ the following theorem holds.

Theorem 3 [5]: The number $m(L) = \left[\frac{n}{4}\right]$ is sufficient and sometimes necessary for an r-cover L of a simple polygon with n vertices.

If the polygon is star-shaped, this number $\left[\frac{n}{4}\right]$ is not economic.
A cover L is said to be a minimum r-cover of a polygon P ($r \in \{p, e, d, t, d^*, t^*\}$) if m(L) is minimum among all r-covers of P.
We state the following problem (MIN-r-cover):

Instance: A polygonal region $P = (v_o, \ldots, v_n)$ and a positive integer exist.
Question: Does an r-cover L of P exist with $m(L) \leq k$?

In 1986, D. T. LEE and A. K. LIN [8] showed that the MIN-e-cover problem is NP-hard. We show the following theorem.
Theorem 4 : For $r \in \{d, d^*, t, t^*\}$ the MIN-r-cover problem is NP-complete.
Proof: We define the Problem 3-SAT ([4]):
3-SAT:
Instance: A set $U = \{u_1, \ldots, u_n\}$ of Boolean variables and a collection $C = \{c_1, \ldots, c_m\}$ of clauses over U exist such that $c_i \in C$ is a disjunction of precisely three literals.

Question: Does a satisfying truth assignment exist such that the conjunctive normal form $c_1c_2...c_m$ is true.

Lemma 1 [4]: The problem 3-SAT is NP-complete.

Given an instance of 3-SAT we construct a simple polygon in polynomial time such that P is coverable by k or fewer "r-pieces" of P if and only if the given instance of 3-SAT is satisfiable. For the details see [5].

Lemma 2: 3-SAT is polynomial reducible to MIN-r-cover.

The theorem 4 follows from Lemma 1 and Lemma 2.

4. Open problems

For covers with triangles, whose sides are diagonals and not edges of the polygon the analogous problem MIN-cover is also NP-complete.

J. SACK [11] has studied so-called rectilinear polygons. A rectilinear polygon is one whose edges are all aligned with a pair of orthogonal coordinate axes. SACK gives an elegant proof for the following theorem of KAHN, KLAWE and KLEITMAN ([6]).

Theorem 5: $\lfloor\frac{n}{4}\rfloor$ guards are occasionally necessary and always sufficient to cover a rectilinear polygon with n vertices.

For many polygons this number is not economic. We state the following open problems:

Problem 1: Is the problem MIN-r-cover for rectilinear polygons also NP-hard?

Problem 2: What is the complexity of regular decomposing a rectilinear polygon into the minmum number of star-shaped components?

Literature

1 Aggarwal, A.: The art gallery theorem: its variations, applications and algorithmic aspects. Ph. D. thesis, Johns Hopkins Univ. 1984.

2 Chvátal, V.: A combinatorial theorem in plane geometry, J.Combin. Theory Ser. B 18 (1975), 39-41.

3 Fisk, S.: A short proof of Chvátals watchman theorem, J. Combin. Theory Ser. B 24 (1978), 374.

4 Garey, M. R., and D. S. Johnson: Computers and Intractability. A Guide to the Theory of NP-Completeness. San Francisco 1979.

5 Hecker, H.-D., and D. Herwig: Some NP-hard polygon-cover problems. To be published.

6 Kahn, J.,, M. Klawe, and D. Kleitman: Traditional galleries require fewer watchman. SIAM J. Alg. Disc. Meth. 4 (1983), 194-206.

7 Keil, M.: Decomposing a polygon into simpler components. SIAM J. of Computing, vol. 14, No. 4 (1985), 799-817.

8 Lee, D. T., and A.K. Lin: Computational complexity of art gallery problems. IEEE Transactions Inf. Theory, IT-32 (1986), 276-282.

9 O'Rourke, J.: Galleries need fewer mobile guards: a variation on Chvátals theorem. Geometriae Dedicata 14 (1983), 273-283.

10 O'Rourke, J.: Art Gallery Theorems and Algorithms. Oxford University press 1987.

11 Sack, J.: Rectilinear computational geometry. Carleton Univ. tech. report SCS-TR-54 (1984).

Translative Integral Geometry

Wolfgang Weil [1)]

1. Introduction

Classical integral geometry is closely related to estimation problems in Stereology and Image Analysis. Most stereological formulae, in particular the *"Fundamental Formulae of Stereology"* have been shown to be variants of integral geometric results (see [8], [19], [27], [28]). The Crofton Formula and the Principal Kinematic Formula are the best known and most used of these results but in the last years several further integral geometric identities have found stereological interpretations which made them rather useful for applications (e.g. Federer's coarea formula [2] and the Blaschke–Petkantschin formulae [13], [20]).

A typical geometrical problem from Image Analysis occurs if a field X of particles in $\mathbb{R}^d$ (the usual case is $d = 2$) can only be observed inside a bounded sampling window $K' \subset \mathbb{R}^d$. If an observation variable f is considered which depends only on the individual particles of X (more precisely on the part of the particles in K'), the total information on f in K' is

$$I = \sum_{K \in X} f(K \cap K') .$$

From this quantity, information on X can realistically only be inferred if either the sampling window K' is chosen randomly or if X is a random field (a geometric point process). In both cases, I is a random variable the mean of which, $\mathbb{E}\, I$, gives the quantity, for which I is a natural (and unbiased) estimator. If the distribution of K' or X as well as the observation variable f are motion invariant, then

$$\mathbb{E}\, I = \gamma \int_{\mathcal{K}_0} \int_{G_d} f(gK \cap K')\, d\mu(g)\, dP_0(K) . \qquad (1)$$

Here γ is the number intensity of X, $\mathcal{K}_0$ the space of all possible shapes (translation classes of sets), and P_0 the distribution of shapes. Also, G_d is the group of rigid motions with invariant measure μ.

1) Universität Karlsruhe Mathematisches Institut II D–7500 Karlsruhe

A further treatment of (1) thus naturally involves a kinematic formula for $\int_{G_d} f(gK \cap K')\, d\mu(g)$. If X is only stationary and not isotropic (respectively, if the distribution of K' is only translation invariant), then (1) has to be replaced by

$$\mathbb{E}\, I = \gamma \int_{\mathcal{K}_0} \int_{\mathbb{R}^d} f(K_x \cap K')\, d\lambda(x)\, dP_0(K), \tag{2}$$

where the subscript x denotes the translate by x and λ is the Lebesgue measure on $\mathbb{R}^d$. Thus, whenever a stationary non–isotropic random structure underlies a problem in Stereology or Image Analysis, translative integral geometry comes in (as long as it is not possible to work with randomly rotated sampling windows (or sections, etc.)). In particular, translative integral formulae cannot be avoided if overlap problems of anisotropic particle fields are treated. Moreover, the degree of anisotropy of a (random) structure is frequently the phenomenon one is interested in. Then again, translative versions of the classical formulae are necessary.

Although some translative integral formulae in dimensions 2 and 3 have been studied quite early, a systematic treatment started just recently. In the following, a short historical survey on Translative Integral Geometry is given and then some of the recent results are explained together with their possible applications. Details and proofs are omitted, but the appropriate references are given. For simplicity, we restrict our considerations to the class $\mathcal{K}$ of convex bodies (compact convex sets) in $\mathbb{R}^d$, some of the results can be extended to the convex ring.

2. Kinematic and Translative Integral Formulae

The most celebrated result in integral geometry is probably the *Principal Kinematic Formula* which (in general d–dimensional setting) has the form

$$\int_{G_d} V_j(K \cap gK')\, d\mu(g) = \sum_{m=j}^{d} c^{(j)}_{m,d+j-m}\, V_m(K) V_{d+j-m}(K'),\ j = 0,\dots,d. \tag{3}$$

Here K, K' are elements of $\mathcal{K}$, $c^{(j)}_{m,d+j-m}$ are known constants (depending on the normalization of μ), and V_j are the classical quermassintegrals (in a slightly different enumeration) given by the *Steiner Formula* for the volume V_d of the parallel body $K + \varepsilon B$,

$$V_d(K + \varepsilon B) = \sum_{j=0}^{d} \varepsilon^{d-j} \kappa_{d-j} V_j(K) \tag{4}$$

($K \in \mathcal{K}$, $\varepsilon > 0$, B unit ball, "+" vector addition, κ_{d-j} $(d-j)$-volume of the $(d-j)$-dimensional unit ball). The first proofs of (3) for $d = 2$ and $d = 3$ go back to Blaschke and Santaló ([5], [21]), the general case was settled by Chern [7]. Blaschke already formulated (3) for certain nonconvex sets as well. Such extensions were developed further in two different directions by Hadwiger [12] and Federer [9]. (For more details and historical remarks on this classical part of integral geometry, see [12] and [22].) Modern developments concern mostly versions of the Principal Kinematic Formula dealing with the local analogs of the quermassintegrals, the curvature measures of Federer ([9], [23], [34]). With a suitable extension of the quermassintegrals, (3) and its local variant now hold in quite general set classes (Zähle [34]). Also, for convex bodies K,K' rather elementary proofs of (3) are possible ([16] and [26]). In particular, from the proof in [26], the Principal Kinematic Formula immediately implies the *Crofton Formula*

$$\int_{\mathcal{E}_q^d} V_j(K \cap E)\, d\mu_q(E) = c^{(j)}_{q,d+j-q}\, V_{d+j-q}(K),\ j = 0,\dots,q. \tag{5}$$

Here $\mathcal{E}_q^d$ is the homogeneous space of q-flats in $\mathbb{R}^d$ with motion invariant measure μ_q, $q \in \{0,\dots,d\}$. (The projection formulae of Cauchy–Kubota can also be obtained from (3).)

Iterations of (3) are simple, for the k-fold version we get

$$\int_{G_d} \cdots \int_{G_d} V_j(K_0 \cap g_1K_1 \cap \cdots \cap g_kK_k)\, d\mu(g_1) \cdots d\mu(g_k)$$

$$= \sum_{\substack{m_0,\dots,m_k=j,\\ m_0+\cdots+m_k=dk+j}}^{d} c^{(j)}_{m_0,m_1,\dots,m_k} V_{m_0}(K_0)V_{m_1}(K_1)\cdots V_{m_k}(K_k) \tag{6}$$

again with known constants $c^{(j)}_{m_0,m_1,\dots,m_k}$. Such iterations are necessary, e. g., if intersection properties or overlap problems in fields of compact particles are studied ([8], [14], [33], [35]).

Translative versions of (3) and (5) have been studied for $d = 2$ and $d = 3$ almost simultaneously with the kinematic case by Blaschke [6] and Berwald–Varga [3]. For $d = 2$, also an iterated version has been given ([6] and [18]). For general d, the kinematic formulae for the volume and the surface area (V_{d-1} is half the surface area) immediately transfer to the translative case.

Furthermore, a translative formula for $j = 0$ can easily be derived (Groemer [11], see also section 4). The following translative version of (5) is due to Schneider [24],

$$\int_{E^\perp} V_j(K \cap E_x)\, d\lambda_{d-q}(x) = \tilde{c}_{dq}^{(j)}\, V(\underbrace{K,\ldots,K}_{d+j-q},\underbrace{B_q,\ldots,B_q}_{q-j}),\ j = 0,\ldots,q. \tag{7}$$

(here $E \in \mathcal{E}_q^d$ is a fixed subspace with orthogonal space $E^\perp$, λ_{d-q} the Lebesgue measure in $E^\perp$, B_q the unit ball in E, and the right–hand side of (7) is a mixed volume (see section 3)).

From general results of McMullen on translation invariant valuations (see the survey [17]) it follows that

$$\int_{\mathbb{R}^d} V_j(K \cap K'_x)\, d\lambda(x) = V_j(K) + \sum_{k=j+1}^{d-1} \Phi_{k,d+j-k}^{(j)}(K,K') + V_j(K'), \tag{8}$$

where $\Phi_{k,d+j-k}^{(j)}(K,K')$ is a mixed valuation depending homogeneously of degree k on K and homogeneously of degree $d+j-k$ on K'. A detailed discussion of (8) is found in [10] and [26]. For a translative analog of (6), (8) has to be iterated. With each iteration, new mixed valuations occur. The corresponding expression (obtained in [31]) is

$$\int_{\mathbb{R}^d} \cdots \int_{\mathbb{R}^d} V_j(K_0 \cap (K_1+x_1) \cap \cdots \cap (K_k+x_k))\, d\lambda(x_1) \cdots d\lambda(x_k)$$

$$= \sum_{\substack{m_0,\ldots,m_k=j,\\ m_0+\cdots+m_k=dk+j}}^{d} \Phi_{m_0,m_1,\ldots,m_k}^{(j)}(K_0,K_1,\ldots,K_k). \tag{9}$$

This introduces a variety of mixed functionals $\Phi_{m_0,m_1,\ldots,m_k}^{(j)}$ defined on the product space $\underbrace{\mathcal{K}\times\cdots\times\mathcal{K}}_{k+1}$, $j \in \{0,\ldots,d\}$, $k = 0,1,\ldots$. Since $\Phi_{m_0}^{(j)} = \Phi_j^{(j)} = V_j$, for $k = 0$, these functionals extend the quermassintegrals. Some of the basic properties of $\Phi_{m_0,m_1,\ldots,m_k}^{(j)}$ (obtained in [26] and [31]) are

(i) continuity (in each argument),

(ii) symmetry:

$$\Phi_{m_{\pi(0)},m_{\pi(1)},\ldots,m_{\pi(k)}}^{(j)}(K_{\pi(0)},K_{\pi(1)},\ldots,K_{\pi(k)}) = \Phi_{m_0,m_1,\ldots,m_k}^{(j)}(K_0,K_1,\ldots,K_k)$$

(for each permutation π of $(0,\ldots,k)$),

(iii) additivity:

$$\Phi^{(j)}_{m_0,m_1,\ldots,m_k}(K_0 \cup K_0', K_1, \ldots, K_k) + \Phi^{(j)}_{m_0,m_1,\ldots,m_k}(K_0 \cap K_0', K_1, \ldots, K_k)$$

$$= \Phi^{(j)}_{m_0,m_1,\ldots,m_k}(K_0, K_1, \ldots, K_k) + \Phi^{(j)}_{m_0,m_1,\ldots,m_k}(K_0', K_1, \ldots, K_k)$$

(if $K_0 \cup K_0' \in \mathscr{K}$),

(iv) homogeneity:

$$\Phi^{(j)}_{m_0,m_1,\ldots,m_k}(\alpha K_0, K_1, \ldots, K_k) = \alpha^{m_0}\, \Phi^{(j)}_{m_0,m_1,\ldots,m_k}(K_0, K_1, \ldots, K_k)$$

(for $\alpha \geq 0$),

(v) decomposability:

$$\Phi^{(j)}_{m_0,m_1,\ldots,m_{k-1},d}(K_0, K_1, \ldots, K_{k-1}, K_k) = \Phi^{(j)}_{m_0,m_1,\ldots,m_{k-1}}(K_0, K_1, \ldots, K_{k-1}) \cdot V_d(K_k).$$

In particular, (v) shows that, in view of the condition $m_0 + \cdots + m_k = dk + j$, the variety of functionals in (9) can (for fixed $j \in \{0,\ldots,d\}$) be reduced to the finite family

$$\mathscr{F}_j = \{\, \Phi^{(j)}_{m_1,\ldots,m_d} \mid m_1, \ldots, m_d \in \{j,\ldots,d\},\ m_1 + \cdots + m_d = d(d-1) + j \,\}.$$

For the further investigations it would be important to have a geometrical description of these mixed functionals $\Phi^{(j)}_{m_1,\ldots,m_d}$. In two special cases (for polytopes and for smooth centrally symmetric bodies) such a result was obtained in [10] and [31], but the general case is still open. A well explored mixed functional for convex bodies which has properties similar to (i),...,(iv) is the *mixed volume* V. It may therefore be conjectured that there is a connection between $\Phi^{(j)}_{m_1,\ldots,m_d}$ and V. Such a relation would be helpful in particular for applications since inequalities for mixed volumes could be used (see section 5, for an example).

3. Mixed Volumes

The mixed volumes $V(K_1, \ldots, K_d)$ of convex bodies $K_1, \ldots, K_d$ are defined as the coefficients in the polynomial expansion which generalizes Steiner's formula (4)

$$V_d(\varepsilon_1 K_1 + \cdots + \varepsilon_k K_k) = \sum_{i_1,\ldots,i_d=1}^{k} \varepsilon_{i_1}\cdots\varepsilon_{i_d}\, V(K_{i_1},\ldots,K_{i_d}) \tag{10}$$

($K_i \in \mathcal{K}$, $\varepsilon_i > 0$, $i=1,\ldots,k$). The quermassintegrals $V_j(K)$ are (up to a normalizing constant) the mixed volumes $V(\underbrace{K,\ldots,K}_{j},\underbrace{B,\ldots,B}_{d-j})$, $j=0,\ldots,d$. The theory of mixed volumes was initiated by Minkowski and pushed further by Aleksandrov (see [15]). The most powerful tools in this theory are the Aleksandrov–Fenchel inequalities which generalize many of the classical inequalities for quermassintegrals.

Comparing the mixed functionals $\Phi^{(j)}_{m_1,\ldots,m_d}$ and V one notices that the degrees of homogeneity are different. The total degree of homogeneity of $\Phi^{(j)}_{m_1,\ldots,m_d}$ (i.e. the sum $m_1 + \cdots + m_d$) is $d(d-1)+j$, whereas V is homogeneous of degree 1 in each argument, hence the total degree of V is d. Even if (v) is used to reduce $\Phi^{(j)}_{m_1,\ldots,m_d}$ to $\Phi^{(j)}_{m_1,\ldots,m_k}$ with $k < d$, the total degree is $d(k-1)+j$. Thus, besides the trivial case $k=1$, a direct connection between the two mixed functionals is only possible if $k=2$ and $j=0$. In this case,

$$\Phi^{(0)}_{m,d-m}(K,K') = \binom{d}{m} V(\underbrace{K,\ldots,K}_{m},\underbrace{-K',\ldots,-K'}_{d-m}), \tag{11}$$

where $-K'$ is the set K' reflected in the origin O. This is the result of Groemer [11] mentioned earlier.

In general, there are two possibilities to enlarge the (total) degree of homogeneity of mixed volumes: (a) by a Crofton–type integral; however, there are no classical formulas for such integrals of mixed volumes,

(b) by using non-linear transformations of the bodies $K_1,\ldots,K_d$. A familiar procedure of this type attachs to each body K its *projection body* ΠK. (ΠK is the body centred at O with support function $h(\Pi K,u)$ given by the $(d-1)$–volume of the projection of K onto the hyperplane orthogonal to u, $\|u\| = 1$; see [25], for more information on projection bodies.) Since $\Pi\alpha K = \alpha^{d-1}\Pi K$, $\alpha > 0$, this transformation increases the degree of homogeneity.

Up to now, there are only some partial results in these directions, a general principle is still not evident. Concerning method (a), it was shown in [10] that

$$\Phi^{(j)}_{m,d+j-m}(K,K') \doteq \int_{\mathcal{E}^d_{d-j}} V(\underbrace{K\cap E,\ldots,K\cap E}_{m-j},\underbrace{-K',\ldots,-K'}_{d+j-m})\, d\mu_{d-j}(E), \tag{12}$$

$m=j+1,\ldots,d-1$. (Here, and in the following, $\doteq$ means "up to a known constant".) For $j = 0$,

(12) reduces to (11). With respect to (b), it is shown in [31] that for centrally symmetric $K_1,\dots,K_{d-j}$

$$\Phi^{(j)}_{d-1,\dots,d-1}(K_1,\dots,K_{d-j}) \doteq V(\Pi K_1,\dots,\Pi K_{d-j},\underbrace{B,\dots,B}_{j}). \tag{13}$$

In both results, $j \in \{0,\dots,d-1\}$ is arbitrary. It is not clear whether (13) holds for arbitrary bodies $K_1,\dots,K_{d-j}$ or whether (12) can be extended to functionals $\Phi^{(j)}_{m_1,\dots,m_k}$ by using multiple integrals.

The situation is simpler in dimensions 2 and 3, which are the important ones for applications. In dimension 2, the only mixed functional occuring (besides V_j) is $\Phi^{(0)}_{1,1}(K,K')$ which is proportional to the *mixed area*,

$$\Phi^{(0)}_{1,1}(K,K') = 2A(K,-K')$$

(this follows from (11) but was already obtained in [6]). In dimension 3, similarly,

$$\Phi^{(0)}_{1,2}(K,K') = 3V(K,-K',-K'),$$

the other mixed functionals are $\Phi^{(0)}_{2,2,2}(K,K',K'')$ and $\Phi^{(1)}_{2,2}(K,K')$. Here (under the assumption of central symmetry) (13) implies

$$\Phi^{(0)}_{2,2,2}(K,K',K'') = \tfrac{3}{4}V(\Pi K,\Pi K',\Pi K'')$$

and

$$\Phi^{(1)}_{2,2}(K,K') = \tfrac{3}{4}V(\Pi K,\Pi K',B).$$

4. Integral Formulae for Mixed Functionals and Mixed Volumes

Since the mixed functionals $\Phi^{(j)}_{m_0,m_1,\dots,m_k}$ were obtained from an (iterated) integral geometric formula, it is immediate that they obey more integral relations. In particular, we have

$$\int_{\mathbb{R}^d} \Phi^{(j)}_{m_0,m_1,\dots,m_{k-2},m}(K_0,K_1,\dots,K_{k-2},K_{k-1}\cap (K_k)_x)\, d\lambda(x) = \sum_{\substack{m_{k-1},\, m_k = m,\\ m_{k-1}+m_k = d+m}}^{d} \Phi^{(j)}_{m_0,m_1,\dots,m_k}(K_0,K_1,\dots,K_k). \tag{14}$$

(which follows directly from (9)) and

$$\int_{SO_d} \cdots \int_{SO_d} \Phi^{(j)}_{m_0,m_1,\ldots,m_k}(K_0, \vartheta_1 K_1, \ldots, \vartheta_k K_k)\, d\nu(\vartheta_1) \cdots d\nu(\vartheta_k)$$

$$= c^{(j)}_{m_0,m_1,\ldots,m_k} V_{m_0}(K_0) V_{m_1}(K_1) \cdots V_{m_k}(K_k), \tag{15}$$

where SO_d is the rotation group with invariant measure ν, $\nu(SO_d) = 1$. (15) follows by comparison of (9) with (6). Moreover, there is a *Crofton Formula* ([31])

$$\int_{\mathscr{E}_q^d} \Phi^{(j)}_{m_0,m_1,\ldots,m_k}(K_0 \cap E, K_1, \ldots, K_k)\, d\mu_{q-j}(E)$$

$$= c^{(j)}_{q,d+j-q} \Phi^{(d+j-q)}_{m_0+d-q,m_1,\ldots,m_k}(K_0, K_1, \ldots, K_k). \tag{16}$$

Integral geometric formulae for mixed volumes have been known only for the quermassintegrals. The above results can be used in reverse to start what one may call an integral geometry for mixed volumes. More precisely, we get the following two results of Crofton type (these and other relations are proved in [10] and [31]),

$$\int_{\mathscr{E}_j^d} V(\underbrace{K \cap E, \ldots, K \cap E}_{j}, \underbrace{K', \ldots, K'}_{d-j})\, d\mu_j(E) \doteq V_d(K) V_{d-j}(K'),\ j=1,\ldots,d, \tag{17}$$

and

$$\int_{\mathscr{E}_{j+1}^d} \int_{\mathscr{E}_{d-j+1}^d} V(\underbrace{K \cap E, \ldots, K \cap E}_{j}, \underbrace{K' \cap F, \ldots, K' \cap F}_{d-j})\, d\mu_{j+1}(E)\, d\mu_{d-j+1}(F)$$

$$\doteq V(\Pi K, \Pi L, B, \ldots, B),\ j=1,\ldots,d-1. \tag{18}$$

In particular, for $j=1$, (18) implies an interesting Crofton formula for support functions ([32])

$$\int_{\mathcal{E}_2^d} h(K \cap E, \cdot)\, d\mu_2(E) \doteq Rh(\Pi K, \cdot)\,. \tag{19}$$

Here R is the spherical Radon transform.

5. Applications

In the introduction, we have mentioned some problems in Image Analysis and Stereology to motivate the need for translative integral formulae. After the description of the present state of Translative Integral Geometry we therefore want to discuss two applications. Further examples, in particular for the formulae (7) and (8) are mentioned in [29].

The first application concerns an overlap problem for random convex particles. If a field of particles in $\mathbb{R}^2$ can be modelled by a stationary and isotropic Poisson process X on $\mathcal{K}$, then the quermass densities of the union set Y can be expressed by those of X. More precisely, if we write A_A, L_A and χ_A for the densities of the area, the boundary length and the Euler characteristic of Y, γ for the intensity (number density) of X, and $A(X)$, $L(X)$ for the mean of the area and the boundary length w.r.t. the shape distribution P_0, then

$$A_A = 1 - e^{-\gamma A(X)}, \tag{20a}$$

$$L_A = e^{-\gamma A(X)} \gamma L(X)\,, \tag{20b}$$

$$\chi_A = e^{-\gamma A(X)} (\gamma - \tfrac{1}{4\pi}\gamma^2 L^2(X))\,. \tag{20c}$$

This result is well–known (see [30], for references), a proof can be based on (6). From observations on Y, the densities A_A, L_A and χ_A can be estimated, (20) then gives the possibility to estimate the corresponding quantities of X, in particular the (mean) particle number γ. If X is not isotropic, (9) has to be used. (20a) and (20b) remain unchanged, instead of (20c) we get

$$\chi_A = e^{-\gamma A(X)} (\gamma - \gamma^2 A(X,-X)) \tag{21}$$

where $A(X,-X)$ is the mean mixed area w.r.t. $P \otimes P$. Introducing a *mean body* $\mathcal{M}(X)$ of X by

$$h(\mathcal{M}(X), \cdot) = \int_{\mathcal{K}_0} h(K, \cdot)\, dP_0(K) \tag{22}$$

we have $A(X,-X) = A(\mathcal{M}(X),-\mathcal{M}(X))$ and $L(X) = L(\mathcal{M}(X))$. If P_0 is symmetric (w.r.t. reflections in O), then $\mathcal{M}(X) = -\mathcal{M}(X)$, hence $A(\mathcal{M}(X),-\mathcal{M}(X)) = A(\mathcal{M}(X))$ and an application of the *Isoperimetric Inequality* gives

$$\chi_A \geq e^{-\gamma A(X)}(\gamma - \tfrac{1}{4\pi}\gamma^2 L^2(X)) \tag{23}$$

with equality for isotropic X (but not only). Details can be found in [30] together with a discussion of the three dimensional case. For non-symmetric P_0 we have $A(\mathcal{M}(X),-\mathcal{M}(X)) \leq \frac{\sqrt{3}}{18} L^2(\mathcal{M}(X))$ by a recently proved inequality ([4]) and thus

$$\chi_A \geq e^{-\gamma A(X)}(\gamma - \tfrac{\sqrt{3}}{18}\gamma^2 L^2(X)) \tag{24}$$

with equality if and only if X consists of congruent equilateral triangles. (23) and (24) can be used (together with (20a) and (20b)) to obtain upper estimates for the particle number γ.

The second application arises from (19). Consider independent IUR-sections $K \cap E_1, \ldots, K \cap E_n$ (i.e. random planar sections with uniform isotropic distribution) of a convex particle $K \subset \mathbb{R}^3$, which are to be used to get some information on the shape of K. Usually, it is possible to determine the directions of the sectioning planes $E_1, \ldots, E_n$, but it is difficult to handle the position of the sections relative to each other. The convex hull of $(K \cap E_1) \cup \cdots \cup (K \cap E_n)$ is therefore not a suitable estimate for K. However, the shape of the vector sum

$$K_n = \frac{1}{n}((K \cap E_1) + \cdots + (K \cap E_n))$$

is invariant against translations of the individual sections. For $n \to \infty$, K_n converges (almost surely, by the law of large numbers for random sets [1]) to the *intersection mean body* $\mathcal{M}_2(K)$ given by

$$h(\mathcal{M}_2(K),\cdot) = \int_{\mathcal{E}_2^3} h(K \cap E,\cdot)\, d\mu_2(E). \tag{25}$$

In case K is centrally symmetric, (19) shows that $\mathcal{M}_2(K)$ uniquely determines K (up to a translation), such that it is possible (in principle) to obtain the shape of K asymptotically from the sections $K \cap E_1, \ldots, K \cap E_n$. Of course, there are numerical problems related to the inversion of the Radon transform R and of the operator Π. For linear sections a corresponding result does <u>not</u> hold, since the intersection mean body $\mathcal{M}_1(K)$ (defined similarly to (25)) is always a ball ([32]).

This concept immediately extends to a stationary particle process X on $\mathcal{K}$ with symmetric particles. Using the mean body introduced in (22) and a similarly defined mean projection body $\mathcal{M}(\Pi X)$ of X (where, in (22), K is replaced by ΠK), we get

$$\int_{\mathcal{L}_2^3} h(\mathcal{M}(X \cap E),\cdot)\, d\nu_2(E) \doteq Rh(\mathcal{M}(\Pi X),\cdot)\,, \tag{26}$$

($\mathcal{L}_2^3$ is the set of planes (in $\mathbb{R}^3$) through O, and ν_2 the invariant measure on $\mathcal{L}_2^3$, $\nu_2(\mathcal{L}_2^3) = 1$.) Hence, from observations of $X \cap E$ in planes E with random directions the mean projection body $\mathcal{M}(\Pi X)$ can be estimated. Since the geometric behaviour of projection bodies is well-known, (26) may be used to obtain information about the anisotropy of the particle system X (more details will appear in [32]).

LITERATURE

[1] Artstein, Z., Vitale, R.A.: A strong law of large numbers for random compact sets. *Ann. Probability* 5 (1975), 879–882.

[2] Baddeley, A.J.: Vertical sections. In *Stochastic Geometry, Geometric Statistics, Stereology*, eds. R.V. Ambartzumian and W. Weil, Teubner, Leipzig 1984, 43–52.

[3] Berwald, L., Varga, O.: Über die Schiebung im Raum. *Math. Z.* **42** (1937), 710–736.

[4] Betke, U., Weil, W.: An isoperimetric inequality for the mixed area of plane convex sets. (In preparation).

[5] Blaschke, W.: Integralgeometrie 13. Zur Kinematik. *Math. Z.* **41** (1936), 465–478.

[6] Blaschke, W.: Integralgeometrie 21. Über Schiebungen. *Math. Z.* **42** (1937), 339–410.

[7] Chern, S.S.: On the kinematic formula in the euclidean space of n dimensions. *Amer. J. Math.* **74** (1952), 227–236.

[8] Davy, P.: *Stereology – A Statistical Viewpoint*. Thesis, Australian National Univ., Canberra 1978.

[9] Federer, H.: Curvature measures. *Trans Amer. Math. Soc.* **93** (1959), 418–491.

[10] Goodey, P., Weil, W.: Translative integral formulae for convex bodies. *Aequationes Mathematicae* **34** (1987), 64–77.

[11] Groemer, H.: On translative integral geometry. *Arch. Math.* **29** (1977), 324–330.

[12] Hadwiger, H.: *Vorlesungen über Inhalt, Oberfläche und Isoperimetrie*. Springer, Berlin-Göttingen–Heidelberg 1980.

[13] Jensen, E.B., Gundersen, H.J.G.: Fundamental stereological formulae based on isotropically oriented probes through fixed points – with applications to particle analysis. *J. Microsc.* (to appear).

[14] Kellerer, H.G.: Minkowski functionals of Poisson processes. *Z. Wahrscheinlichkeitstheorie verw. Gebiete* **67** (1984), 63–84.

[15] Leichtweiß, K.: *Konvexe Mengen*. Springer, Berlin–Heidelberg–New York 1980.

[16] Mani–Levitska, P.: A simple proof of the kinematic formula. *Mh. Math.* **105** (1988), 279–285.

[17] McMullen, P., Schneider, R.: Valuations on convex bodies. In *Convexity and Its Applications*, eds. P. Gruber and J.M. Wills, Birkhäuser, Basel 1983, 170–247.

[18] Miles, R.E.: The fundamental formula of Blaschke in integral geometry and geometrical probability, and its iteration, for domains with fixed orientations. *Austral. J. Statist.* **16** (1974), 111–118.

[19] Miles, R.E.: The importance of proper model specification in stereology. In *Geometrical Probability and Biological Structures*, eds. R.E. Miles and J. Serra, Springer, Berlin–Heidelberg–New York 1978, 115–136.

[20] Miles, R.E.: A comprehensive set of stereological formulae for embedded aggregates of not–necessarily–convex particles. *J. Microsc.* **138** (1985), 115–125.

[21] Santaló, L.A.: Geometria integral 4. Sobre la medida cinemática en el plano. *Abh. Math. Sem. Univ. Hamburg* **11** (1936), 222–236.

[22] Santaló, L.A.: *Integral Geometry and Geometric Probability*. Addison–Wesley, Reading, Mass. 1976.

[23] Schneider, R.: Parallelmengen mit Vielfachheit und Steiner–Formeln. *Geom. Ded.* **9** (1980), 111–127.

[24] Schneider, R.: Crofton's formula generalized to projected thick sections. *Rend. Circ. Mat. Palermo, Ser. II,* **30** (1981), 157–160.

[25] Schneider, R., Weil, W.: Zonoids and related topics. In *Convexity and Its Applications,* eds. P. Gruber and J.M. Wills, Birkhäuser, Basel 1983, 296–317.

[26] Schneider, R., Weil, W.: Translative and kinematic integral formulae for curvature measures. *Math. Nachr.* **129** (1986), 67–80.

[27] Stoyan, D., Kendall, W.S., Mecke, J.: *Stochastic Geometry and Its Applications*. Akademie-Verlag, Berlin 1987.

[28] Weil, W.: Stereology – a survey for geometers. In *Convexity and Its Applications,* eds. P. Gruber and J.M. Wills, Birkhäuser, Basel 1983, 360–412.

[29] Weil, W.: Point processes of cylinders, particles and flats. *Acta Applicandae Mathematicae* **9** (1987), 103–136.

[30] Weil, W.: Expectation formulas and isoperimetric properties for non–isotropic Boolean models. *J. Microsc.* **150** (1988), (in print).

[31] Weil, W.: Iterations of translative integral formulae and nonisotropic Poisson processes of particles. (Submitted).

[32] Weil, W.: The determination of convex bodies from the mean of random sections. (In preparation).

[33] Weil, W., Wieacker, J.A.: Densities for stationary random sets and point processes. *Adv. Appl. Prob.* **16** (1984), 324–346.

[34] Zähle, M.: Curvature measures and random sets I. *Math. Nachr.* **119** (1984), 327–339.

[35] Zähle, M.: Curvature measures and random sets II. *Probab. Th. Rel. Fields* **71** (1986), 37–58.

Object Recognition in Images using Models of Geometric Structures

B.D. Ripley [1)]

1. Overview

Problems which involve object recognition in images occur in a wide variety of disciplines. Perhaps the most obvious are those of *computer vision* ([11]), which traditionally use edge-detection algorithms to produce a skeleton image followed by clustering algorithms to select objects, and *manufacturing inspection*, discussed in more detail below. Examples from scientific disciplines include the classification of galaxies in images from optical telescopes ([7]) and of biological organisms from micrographs. These tend to be rather subtler than the engineering examples in that identification uses subtle features of shape and (although not in astronomy) texture.

One of the most promising ways to consider geometrical structures in images is *mathematical morphology* ([5], [12], [13]). This is based on transformations of closed sets and has already been used successfully to summarize the shapes in binary images ([8], [10]). In §3 we apply mathematical morphology to the problem of detecting foreign bodies in a food packing plant.

A modern statistical approach to all imaging problems ([1], [3], [4], [6], [8], [10]) is to build a prior model for the image contents, to model the imaging and noise processes and to base inferences on the posterior distribution of the content given the observed image. Although this approach would seem ideally suited to object recognition, it has almost exclusively been applied to image restoration, the only exception known to me being an artificial example in [3]. In §4 we consider its application to recognition.

2. A brief review of 'mathematical morphology'

In its original form (Serra's *Texture Analyser*) of the mid 1960's) mathematical morphology applied to binary images. Let A denote the 'black' phase of the image, which is assumed to extend far beyond the window W through which it is observed. Let T denote a 'test set' or 'structuring element'. Then the basic transformations are the *erosion* $E_T(A)$ of A by T defined by

$$E_T(A) = \{\mathbf{x} \mid \mathbf{x} + T \subset A\}$$

and the *'dilatation'* $D_T(A)$ of A by T given by

$$D_T(A) = \{\mathbf{x} \mid \mathbf{x} + T \text{ intersects } A\}$$

These are dual transformations in the sense that $D_T(A) = [E_T(A^c)]^c$ where ^{c} denotes complementation. From these we can construct

$$O_T(A) = D_{-T}(E_T(A))$$
$$C_T(A) = E_{-T}(D_T(A))$$

the *opening* and *closing* of A by T respectively. In particular, $O_T(A) \subset A \subset C_T(A)$.

Since A is known only through its intersection with W, these transformations are computable only within a smaller window than W ($E_T(W)$ for erosion and dilatation, $E_{T\oplus -T}(W)$ for opening and closing, [8]); this can be seen in some of the later figures.

The advantages of morphological transformations is that they can be fast (even on general-purpose hardware, [10]), that they are not local and so are rather immune to noise and that they do not depend on segmentation (isolating objects in the image).

1) *Department of Mathematics, University of Strathclyde, 26 Richmond Street, Glasgow G1 1XH, UK.*

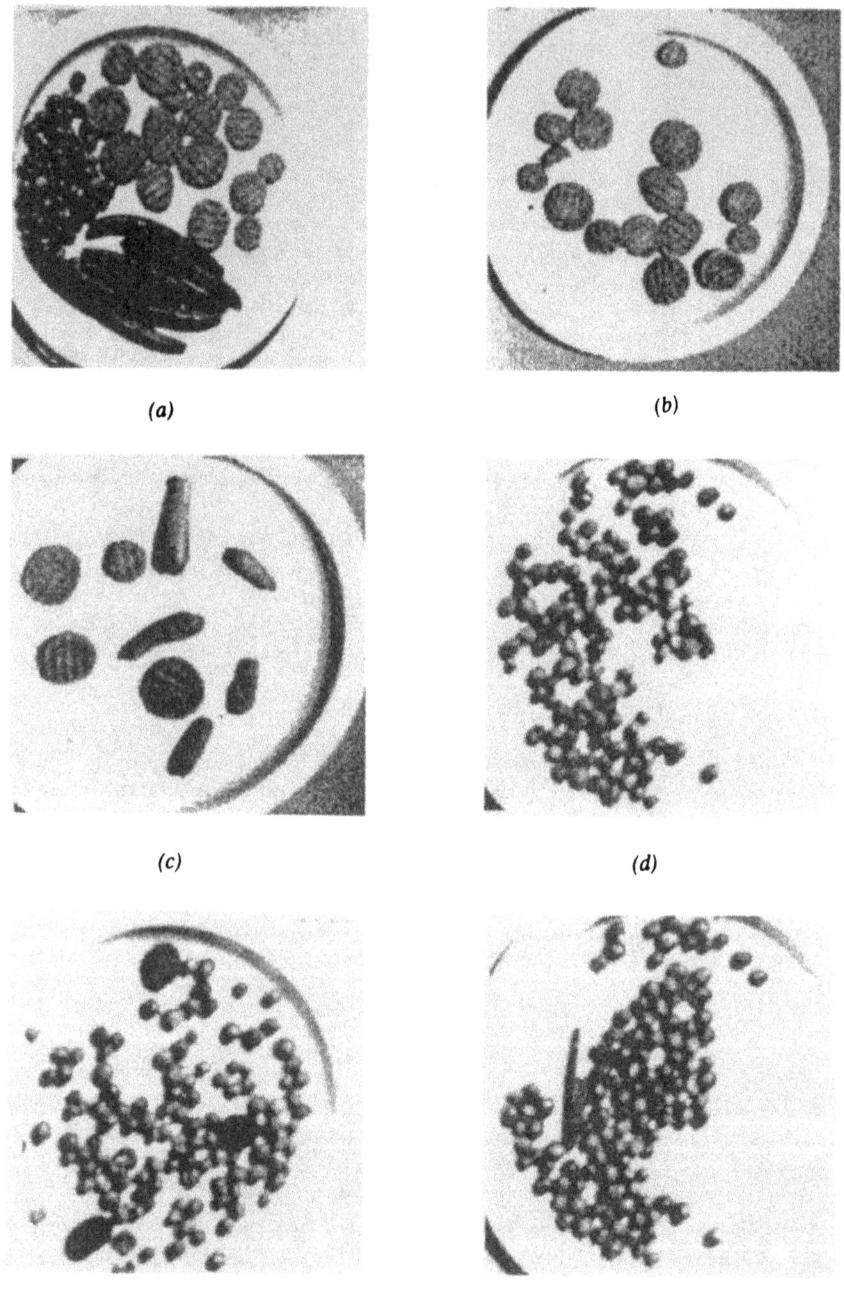

Figure 1. Images of foodstuffs. The originals are at 512×256 resolution with 16 greylevels, obtained by digitizing a TV image with a Haba digitizer and Atari 1040ST microcomputer. *(a)* Peas, beans and cut carrots. *(b)* Cut carrots. *(c)* Cut and uncut carrots. *(d)* Peas alone. *(e)* Peas and olives. *(f)* Peas and a green, shiny, pen top.

3. Filtering out unwanted objects

Figure 1 shows some greylevel images of food, intended to be representative of the machine inspection problem in a frozen food packing plant. From images such as 1(a)–1(d) we are expected to detect the problems in images 1(b,c,e,f) but accept 1(a). These images all have 16 greylevels, and it would be possible to apply greylevel morphology to them directly. We chose, mainly from computational considerations, to filter the images to reduce them to one or more binary images, and to apply binary morphology to the filtered images. To remove noise problems in the (slow-scan) digitizer, the images were reduced to 256×256 pixels at 32 greylevels by averaging horizontally adjacent pixels.

The need for filtering is evident from the observation that humans recognize peas and carrots not only by their colours (suppressed in monochrome images) and intensity, but also by their *textures*. For example, we note that peas are shiny and have a highlight whereas most olives do not, and that cut carrots have a pattern on them from the scalloped-edged cutter. Our overall strategy is to use simple filters to enhance the textures which might well be important, and then to use mathematical morphology to extract the objects which are homogeneous with respect to the desired textures. It is easiest to illustrate this by examples.

Uncut carrots

Image 1(b) contains only carrot slices, one of which is broken. The image was first thresholded and then opened by a disc of radius 8 pixels, the known minimal acceptable radius of a carrot slice. Figure 2(c) show that this is sufficient to isolate the broken slice (and the rim of the plate).

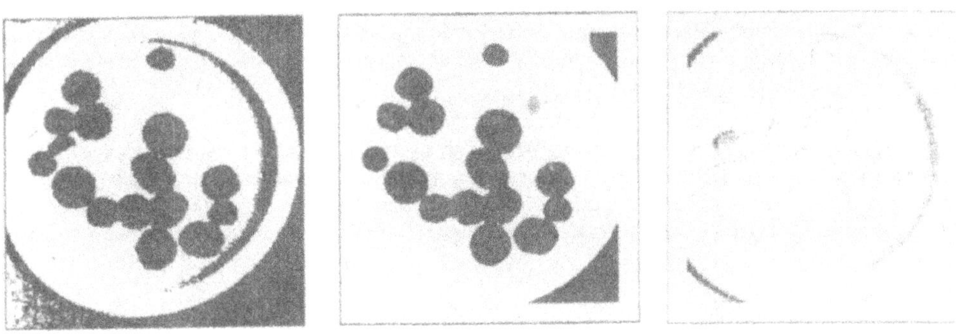

(a) (b) (c)

Figure 3. Transformations of image 1(b). *(a)* Thresholded image, *(b)* opened by a disc of radius 8 pixels. *(c)* Greylevel image masked by part (b).

Cut and whole carrots

Image 1(c) was again thresholded, but in this case some residue of the texture on the carrots remained, so a small closing operation (a disc of radius 2 pixels) was used. At the increased magnification of this figure the appropriate minimum radius is 12 pixels. Once again the sliced carrots, only, are removed from the image on masking.

In both these examples we succeeded in our task using shape alone. It would be better to confirm by texture that the masking images are indeed comprised of cut carrot. As the slices are oriented 'at random' no anisotropic morphological transformation will suffice, but we can use a frequency-domain filter that enhances wavelengths corresponding to that of the cutter, in all directions.

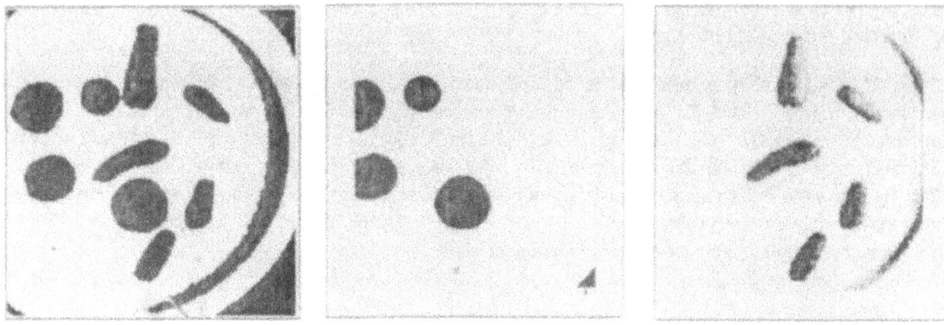

(a) (b) (c)

Figure 4. Transformations of image 1(c). *(a)* Thresholded image, *(b)* opened by a disc of radius 12 pixels. *(c)* Greylevel image masked by part (b).

Peas and foreign bodies

In the previous two examples the possible objects present were known, whereas for peas the task is to label as 'not pea' any object which does not meet the specification of a pea derived from image 1(d), *without* having seen the 'foreign body' before. Using the summaries of binary images developed in [8] and [10] we can see that after thresholding peas will appear as roughly circular objects of diameters between 10 and 15 pixels. Thus any object eliminated by a closing of radius 5 pixels is not a pea, and anything retained by an opening of radius 8 pixels is too big to be a pea. Further, peas have a characteristic texture, a highlight caused by their (uniform) glossiness.

The first example, image 1(e), contains three olives which can be isolated both by size and texture. Size alone (figure 5(a)) is a perfect discriminator. We tried to measure glossiness by a spatial difference filter, the absolute value of the difference of a pixel's luminance and the average of that of its neighbours. This was thresholded and then 'cleaned' by a closing by a disc of radius 4 pixels (figure 5(c)). Note that one of the olives *does* have a highlight, and this is shown in figure 5(c), so the isolation by glossiness is also achieved correctly.

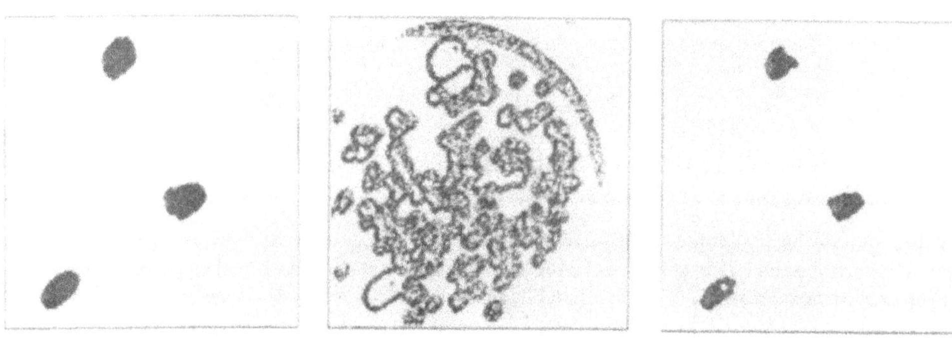

(a) (b) (c)

Figure 5. Transformations of image 1(e). *(a)* Opening by a disc of radius 8 pixels. *(b)* Absolute value of the spatial difference filter. *(c)* Part (b) opened by a disc of radius 4 pixels.

The task of isolating the pen top in figure 1(f) is considerably more difficult any of the previous tasks. The colour, glossiness and curvature are very similar to real peas, and it is quite likely that a human inspector would miss this object. The clip of the pen gives the game away for the shape filter (figure 6(c));.despite several attempts, no useful textural filter was found.

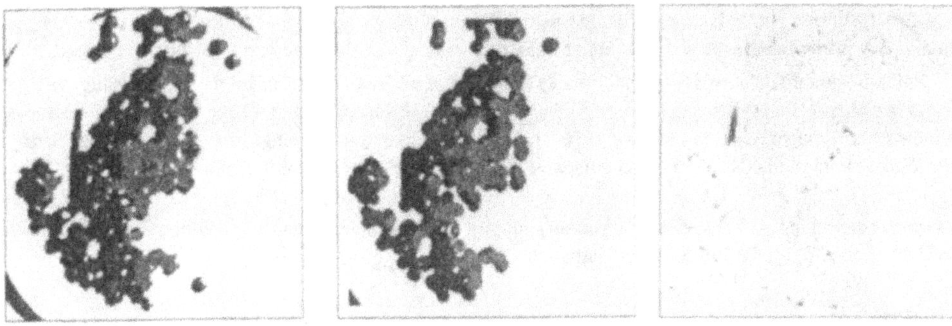

(a) *(b)* *(c)*

Figure 6. Transformations of image 1(f). *(a)* Thresholded image, *(b)* opened by a disc of radius 5 pixels and then dilated by a disc of radius 2 pixels. *(c)* Greylevel image masked by part (b).

4. Models from stochastic geometry

A more formal way to achieve similar aims to those fulfilled in §3 is to build a stochastic prior model of the image contents. The most promising source of candidate models in stochastic geometry (e.g. [14]). The description of the image will be as a point process on a suitable space of objects. In particular, our images of foods could be modelled by 'germ-grain' processes, in which 'centres' of objects are chosen from an (interacting) point process, and the object descriptions are drawn independently of the position. For carrot slices the description would be a (random) radius and the texture, perhaps a sine-wave of fixed frequency and amplitude but uniformly distributed direction.

Perhaps the most useful class of point processes are the so-called Gibbs point process, which are specified by giving an energy U to each configuration of (fully-specified) objects. Then the probability of the configuration is proportional to $\exp(-U)$. Such processes can be simulated quite easily by iterative schemes ([9]).

The formal Bayesian paradigm also needs a model for the distortion and observation process. This will involve integration over pixels (in going from a continuum-based prior to a pixel-based observation) and a model for camera noise. The real problem is to specify the errors which could occur, for these are objects of unknown form. One possibility is to allow a range of small objects which have very low probabilities, and so allow the reconstruction to build fairly general objects from these pieces.

In the first two examples above the prior model could be extended to include whole carrots and broken and under-sized carrot slices. In that case the posterior distribution is over descriptions of the image listing objects, their centres, types and detailed descriptions. A MAP (maximum *a posteriori*) estimate give the most probable overall description. Since the description is finite-dimensional, the maximization can be done numerically once a good starting point is found. Of course, the filtering techniques of §3 provide precisely what is required. Indeed, by varying the filters slightly, it is often the case that a range of starting points will be selected, and precise descriptions can be found by maximizing the posterior probability starting at these descriptions. The maxima obtained then allow a choice of the best fit (measured by posterior probability of the plausible descriptions. This was tried for the carrot examples and works well, producing the visually obvious descriptions from images 1(b) and 1(c).

5. Future directions

Both filters based on mathematical morphology and prior models based on stochastic geometry appear to have a bright future in image processing and analysis. Indeed, it seems that imaging techniques and computational power have at last reached the point that the probability models of stochastic geometry can now be compared with image data in a fully automatic way. This is a very exciting prospect. It will also be possible to build loss functions on reconstructed images using the morphology filters, and so to complete the Bayesian paradigm by minimizing expected loss under a realistic prior given the observation of just one image.

Point process models usually contain (many) parameters whereas Bayesian priors must be fully specified (unless 'empirical Bayes' procedures are to be used, [2]). Thus the parameters of the prior would need to be estimated from images such as 1(a) and 1(d). This is also an exciting challenge, and it appears that Besag's pseudo-likelihood methods (see, in this context, [10], chapter 4) could be very promising.

Acknowledgements This work was partially supported by the Science and Engineering Research Council under its 'Complex Stochastic Systems' initiative.

References

[1] Besag, J. (1986): On the statistical analysis of dirty pictures (with discussion). *J. Roy. Statist. Soc. B* **48**, 259- 302.

[2] Cox, D.R. & Hinkley, D.V. (1974): *Theoretical Statistics*. Chapman & Hall, London.

[3] Grenander, U. (1983): *A Tutorial in Pattern Theory*. Division of Applied Mathematics, Brown University.

[4] Geman, S. & Geman, D. (1984): Stochastic relaxation, Gibbs distributions and the Bayesian restoration of images. *I.E.E.E. Trans. PAMI* **6**, 721-741.

[5] Haralick, R.M., Sternberg, S.R. & Zhuang, X. (1987): Image analysis using mathematical morphology. I.E.E.E. Trans. PAMI **9**, 532-550.

[6] Marroquin, J., Mitter, S. & Poggio, T. (1987): Probabilistic solution of ill-posed problems in computational vision. *J. Amer. Statist. Assoc.* **82**, 76-89.

[7] Murtagh, F. (1988): Classification problems in astronomy. In *'Classification and Related Problems of Data Analysis'* ed. H.H. Bock, Elsevier, 23-32.

[8] Ripley, B.D. (1986): Statistics, images and pattern recognition (with discussion). *Canad. J. Statist.* **14**, 83-111.

[9] Ripley, B.D. (1987): *Stochastic Simulation*. Wiley, New York.

[10] Ripley, B.D. (1988): *Statistical Inference for Spatial Process*. Cambridge University Press, Cambridge.

[11] Rosenfeld, A & Kak, A.C. (1982): *Digital Picture Processing*. Academic Press, Orlando

[12] Serra, J: (1982): *Image Analysis and Mathematical Morphology*. Academic Press, London.

[13] Sternberg, S.R. (1986): Grayscale morphology. *Comput. Vision Graph. Image. Proc.* **35**, 333-355.

[14] Stoyan, D., Kendall, W.S. & Mecke, J. (1987): *Stochastic Geometry*. Akademie-Verlag / Wiley, Berlin / Chichester.

RANDOM PATHS IN NON-CONVEX BODIES

I. Saxl, J. Rataj [1)]

1. Introduction

The theory of random paths is exhaustively elaborated for convex bodies [1 - 3,6,8] and to certain extent also for random closed sets [7,11,12]. Its main results can be useful in the connection with interparticle distances. In this note we shall differentiate between paths - segments σ- connecting two points of the boundary of the examined set and between paths - rays $\hat{\sigma}$- ending or originating in an interior point of the set. The length distributions of segments and rays critically depend on the way in which the randomness of paths arises and describe in a different manner the spatial organization of boundary components.

The following types of random paths are the most important:

μ: the segments σ_μ are generated by IUR (isotropic uniform random) lines hitting the examined set;

ν: segments σ_ν and rays $\hat{\sigma}_\nu$ are generated by a UR point of the interior of the set and an isotropic direction;

ϑ: a ray $\hat{\sigma}_\vartheta$ is the shortest path from a UR point of the interior to the boundary;

γ: a segment σ_γ originates in a UR point of the boundary and has an isotropic direction into the matrix.

2. Non-convex bodies

Let Y be a compact set with a non-empty interior in R^d. Denote by λ_Y the normalized Lebesgue measure on Y and by ω the normalized rotationaly invariant measure on the unit sphere S^{d-1}. We define the measureable mappings from $Y \times S^{d-1}$ to $[0,\infty)$

$$\hat{\sigma}_\nu: (y,u) \longmapsto \sup\{r \geq 0: [y,y+ru] \subseteq Y\},$$
$$\sigma_\nu: (y,u) \longmapsto \hat{\sigma}_\nu(y,u) + \hat{\sigma}_\nu(y,-u).$$

The distribution functions $\hat{F}_\nu$, F_ν of $\hat{\sigma}_\nu$, σ_ν, respectively, are understood with respect to the probability measure $\lambda_Y \otimes \omega$. The mapping

$$\hat{\sigma}_\vartheta: y \longmapsto \sup\{r \geq 0: B(y,r) \subseteq Y\}$$

from Y to $[0,\infty)$ induces the distribution $\hat{F}_\vartheta$ with respect to λ_Y (B(y,r) is the ball of radius r centred in y).

The direct definition of the distribution F_μ of σ_μ is more complicated. However, using the fact that F_ν is length-weighted with respect to F_μ [1,6], we can define [7]

$$F_\mu(dr) = \frac{c}{r} F_\nu(dr) \qquad \text{with} \quad c^{-1} = \int_0^\infty r^{-1} F_\nu(dr) \tag{1}$$

1) Institute of Geophysics, Czechoslovak Academy of Sciences, 141 31 Praha 4 - Spořilov, Boční II/1401, Czechoslovakia

in the case when the last integral is finite (otherwise, the distribution F_μ does not exist as the number of segments generated by an IUR line hitting Y is not finite a.s. This cannot happen if, for example, the boundary ∂Y is (d-1)-rectifiable.)

It has been shown [2,3,11] that all the above introduced length distributions can be expressed by means of the normalized mean volume content of the "overlap body"

$$\Omega(r) = \int \lambda_Y(Y \ominus [0,-ru])\omega(du)$$

and

$$\Omega_B(r) = \lambda_Y(Y \ominus B(0,r)) \ , \quad r \geq 0.$$

We obtain

$$\hat{F}_\nu(r) = 1 - \Omega(r) \ , \tag{2}$$

$$F_\nu(r) = 1 - \Omega(r) + r\Omega'(r) \ , \tag{3}$$

$$\hat{F}_\vartheta(r) = 1 - \Omega_B(r) \ , \tag{4}$$

$$F_\mu(r) = 1 - \Omega'(r)/\Omega'(0) \tag{5}$$

(note that $-\Omega'(0)$ is the volume-normalized mean content of the total projection of Y into the isotropic hyperplanes).

The moments of these distributions depend, in general, on the shape of Y, but there is a notable exception, namely $E\,\sigma_\mu = -1/\Omega'(0)$, which is independent even of the shape of the path [8]. Unfortunately, just this type of the mean free path is commonly used in materials science, as it can be simply estimated by linear analysis of planar sections of the examined specimen [4,5]. Consequently, when used as a material parameter, it should be accompanied by its variance, which is already shape dependent. From (1) we get var $\sigma_\mu = E\,\sigma_\mu(E\,\sigma_\nu - E\,\sigma_\mu)$.

3. Stationary random closed sets and lattices of bodies

If X is a stationary random closed set in R^d with H^{d-1}-rectifiable boundary (for details see [13]), all the free paths introduced above can be defined analogically in the complement $Y = X^c$. If X is represented as a union of isolated subsets (particles), then the notion of the mean interparticle distance becomes clear.

We shall consider two special models. Let Z be a compact set (particle) and X_B the Boolean model with the intensity λ and the primary grain obtained by isotropic rotations of Z. The well known relations are [12]

$$\hat{F}_\nu(r) = 1 - \exp\left\{-\lambda\left[\int \nu^d(Z \oplus [0,-ru])\omega(du) - \nu^d(Z)\right]\right\} \ , \tag{6}$$

$$\hat{F}_\vartheta(r) = 1 - \exp\left\{-\lambda\left[\nu^d(Z \oplus B(0,r)) - \nu^d(Z)\right]\right\} \tag{7}$$

(ν^d is the Lebesgue measure in R^d). $\hat{F}_\nu(r)$ is called the linear contact distribution function, $\hat{F}_\vartheta(r)$ is the spherical contact distribution function. If, moreover, Z is convex, we have $F_\mu = \hat{F}_\nu$ and, applying the Steiner formula, we see that F_μ , $\hat{F}_\nu$ and $\hat{F}_\vartheta$ are functions only of the Minkowski functionals of Z, not depending on the particular shape of Z. In the planar case (d = 2), denoting by $\tilde{a}$, $\tilde{b}$ the area and perimeter of Z, respectively, we obtain in particular

$$E\,\hat{\sigma}_\nu = E\,\sigma_\mu = E\,\sigma_\nu/2 = \delta \quad , \tag{8}$$

$$E\,\hat{\sigma}_\nu^2 = E\,\sigma_\mu^2 = E\,\sigma_\nu^2/3 = 2\,\delta^2 \quad , \tag{9}$$

$$E\,\hat{\sigma}_\vartheta = \exp(w^2)(1-\mathrm{erf}(w))/\sqrt{4\lambda}\ ,\ E\,\hat{\sigma}_\vartheta^2 = (\lambda^{-1} - \tilde{b}E\,\hat{\sigma}_\vartheta)/\pi \quad , \tag{10}$$

where $\delta = \pi/(\lambda\tilde{b})$ and $w = \tilde{b}\sqrt{\lambda/4\pi}$. The formulae (8) - (10) remain true also for a Boolean model with a random convex isotropic primary grain after replacing $\tilde{a}$, $\tilde{b}$ by their mean values [12].

On the other hand, we consider a regular model with the particle Z placed in the vertices of a regular lattice, i.e.

$$X_R = \bigcup \left\{ Z \oplus z.a,\ z \in \mathbb{Z}^d \right\},\quad a \in (0,\infty)^d \quad .$$

In fact, X_R is a deterministic set (a lattice of bodies [9]), but it can be easily made stationary random when applying a UR translation from the cell

$$C_0 = \left\{ x \in R^d:\ 0 \le x^{(i)} \le a^{(i)},\ i=1,\ \dots\ ,d \right\} .$$

As can be easily seen, the free paths $\hat{\sigma}_\nu$, $\hat{\sigma}_\vartheta$ and σ_μ in X_R^c can be computed using only the paths originating in C_0 (the origin of the σ_μ-segment must be defined uniquely according to the orientation of the generating line).

The comparison of the free path distributions in X_B^c and X_R^c seems to be rather interesting and we shall illustrate it by several examples.

4. Examples

As the examples, several planar structures (d = 2) will be considered. All the regular models will be based on the rectangular lattice with the cell $C_0 = \left\{ (x,y):\ 0 \le x \le a,\ 0 \le y \le b \right\}$; we assume that the length ratio $q = b/a \ge 1$. The areal density of particles in lattices as well as in Boolean models will be, unless stated otherwise, $\lambda = 1/(a^2q)$.

4.1. Point structures

Choosing $Z = \{\sigma\}$, we obtain the regular point structure (rectangular lattice of points), in which only the rays $\hat{\sigma}_\vartheta$ are bounded. The distribution function is (introducing $s = 2r/a$)

$$\hat{F}_\vartheta(r) = \frac{\pi s^2}{4q} + \sum_{i=1}^{2}\left[\frac{\alpha_i}{2q}\sqrt{s^2-\alpha_i} - \frac{s^2}{2q}\arccos\frac{\alpha_i}{s}\right]\chi_{[\alpha_i,\alpha_3]}(s)\ ,\ 0 \le s \le \alpha_3\ , \tag{11}$$

where $\alpha_1 = 1,\ \alpha_2 = q,\ \alpha_3 = \sqrt{1+q^2}$, and its moments are

$$E_R\,\hat{\sigma}_\vartheta(\sigma) = a\,H(q),\qquad E_R\,\hat{\sigma}_\vartheta^2(\sigma) = \frac{a^2}{12}(q^2+1)\ , \tag{12}$$

where $H(q) = \frac{1}{12}\left[2\sqrt{1+q^2} + q^{-1}\ln(q+\sqrt{1+q^2}) + q^2\ln(q^{-1}+\sqrt{1+q^{-2}})\right]$.

X_B is the planar Poisson point process in this case and Eq. (10) gives the well known result [12]

$$E_B\,\hat{\sigma}_{\vartheta}(\sigma) = (2\sqrt{\lambda})^{-1}, \qquad E_B\,\hat{\sigma}^2_{\vartheta}(\sigma) = (\pi\lambda)^{-1} \; ; \tag{13}$$

we note that the intensity λ decreases with growing q.

4.2. Particle structures

Particles of two simple shapes, namely the disc $B(\sigma, r_0)$ ($Z = \{(x,y): x^2 + y^2 \leq r_0^2\}$, $2r_0 \leq a$ - lattice of discs) and the rectangle $R(\sigma, c, d)$ ($Z = \{(x,y): |x| \leq c/2, |y| \leq d/2\}$, $c \leq a$, $d \leq b$ - lattice of rectangles) will be considered. The distribution functions of ϑ - rays are only slightly more complicated in comparison with (11) and lead to

$$E_R\,\hat{\sigma}_{\vartheta}(B) = \frac{1}{a^2q - \pi r_0^2}\left[a^3q\,H(q) + \frac{\pi r_0^3}{3} - a^2qr_0\right], \tag{14}$$

$$E_R\,\hat{\sigma}^2_{\vartheta}(B) = \frac{1}{a^2q - \pi r_0^2}\left[-2a^3qr_0\,H(q) - \frac{\pi r_0^4}{6} + a^2qr_0^2 + \frac{a^4}{12}q(q^2+1)\right] \tag{15}$$

and (with $q = (b-d)/(a-c)$)

$$E_R\,\hat{\sigma}_{\vartheta}(R) = \frac{(a-c)^2}{4(ab-cd)}\left[4(a-c)q'\,H(q') + d + q'^2c\right], \tag{16}$$

$$E_R\,\hat{\sigma}^2_{\vartheta}(R) = \frac{(a-c)^3}{12(ab-cd)}\left[aq'^3 + (a-c)q' + d\right] \quad . \tag{17}$$

We have not been able to find the distribution functions of μ - and ν - paths, nevertheless [8]

$$E_R\,\sigma_{\mu}(Z) = \frac{\pi(1-\lambda\tilde{a})}{\lambda\tilde{b}} \tag{18}$$

independently of the particle positions. It is also clear that $E_R\,\hat{\sigma}_{\nu}(Z) = \infty$ (unless $2r_0 = a = b$ or $b = d$ and $a = c$), since equidistant parallel unbounded strips of constant breadth, yielding the infinite mean length of the $\hat{\sigma}_{\nu}$-path [10], can be inscribed into X_R^c. Consequently, also $E\,\sigma^2_{\mu}(Z) = E\,\sigma_{\nu}(Z)\,E\,\sigma_{\mu}(Z) = \infty$.

In the Boolean model X_B, the moments of all paths are given by Eq'ns (9) - (11). We note that $E_B\,\sigma_{\mu}(Z) = \pi/(\lambda\tilde{b}) = \pi(1 - A_A)/\tilde{B}_A$ in the analogy to Eq. (18) ($A_A = 1 - \exp(-\lambda\tilde{a})$ is the areal function of the Boolean model and $\tilde{B}_A$ is its total perimeter per unit area [12]).

4.3. Fibre structures

The most simple fibre structures, namely the lattice of parallel line segments and the corresponding isotropic Boolean model, will be obtained by setting either c=0 or d=0, e.g.
$S(\sigma, c) = Z = \{(x,y): |x| \leq c/2, |y| = 0\}$, in the above described case

of rectangles R. The limit cases of these structures are lattice of points and Poisson point process at $c = d = 0$ and, on the other hand, parallel equidistant lines and stationary isotropic Boolean segment process at, for example, $c = a$, $d = 0$.

Another fibre structure will be obtained with $Z = \{(x,y): 0 \leq x \leq a,\ y = 0 \text{ or } x = 0,\ 0 \leq y \leq b\}$. The regular model ($X_R$) is the rectangular net and the distribution functions and their moments are those of an isolated rectangle derived by Coleman [1] for μ- and ν-paths; the corresponding formulae for ϑ-rays are in [10]. The Boolean equivalent of this structure cannot be calculated by means of Eq'ns (8) - (10) as Z is not convex in this case. Therefore, we compare X_R with the Boolean fibre process with the intensity of the underlying Poisson process $\lambda = 2/a^2q$ and the isotropic primary grain of two lengths, a and b (with equal probability). The detailed results are in [10], where also the asymptotic formulae $q \to \infty$ for μ-, ν-segments and ϑ-rays are given. The limit situation $q \to \infty$ describes again the structure formed by equidistant parallel lines or a stationary isotropic Poisson line process of intensity $L_A = 1/a$.

The numerical results for discs and segments are shown in Fig's 1 and 2. The rays $\hat{\sigma}_\vartheta$ are shorter in the regular model X_R than in the

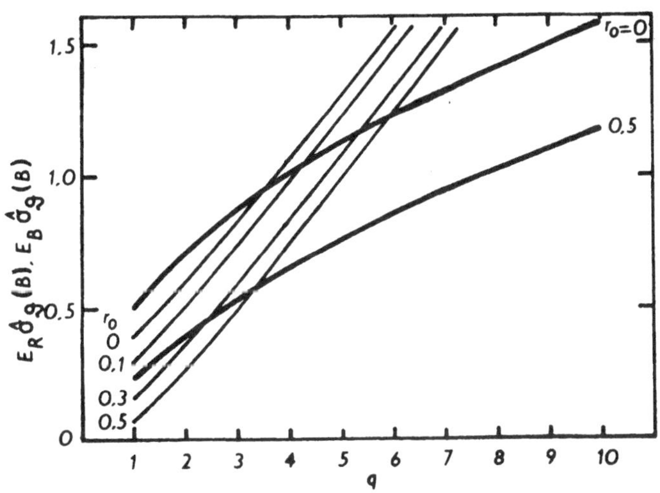

Fig. 1 The dependence of $E_R\hat{\sigma}_\vartheta(B)$ (thin lines) and $E_B\hat{\sigma}_\vartheta(B)$ (thick lines) on the lattice ratio $q = b/a$ for different values of the disc diameter $0 \leq r_0 \leq a/2$; $a = 1$.

Boolean model X_B in the dense quadratical structure; with the increasing edge ratio q this difference decreases. It vanishes at $q = 3.5$ and at even higher values of q, the $\hat{\sigma}_\vartheta$-rays are shorter in the Boolean models. These properties follow directly from the asymptotic behaviour of the models, which will be described in what follows.

Regular models: As $q \to \infty$, the function H(q) approaches q/4, hence by (14) $E_R\hat{\sigma}_\vartheta(B) \to aq/4 - r_0$ and $E_R\hat{\sigma}^2_\vartheta(B) \to a^2q^2/12$. Clearly, from the point of view of $\hat{\sigma}_\vartheta$, the structures approach the point lattice or, equivalently, the parallel lines at the distance b. The rectangles behave similarly unless d is comparable with b. The asymptotic case then approaches an infinite strip of breadth (a-c) with $E_R\,\hat{\sigma}_\vartheta(R) \to (a-c)/4$ and $E_R\,\hat{\sigma}^2_\vartheta(R) \to (a-c)^2/12$.

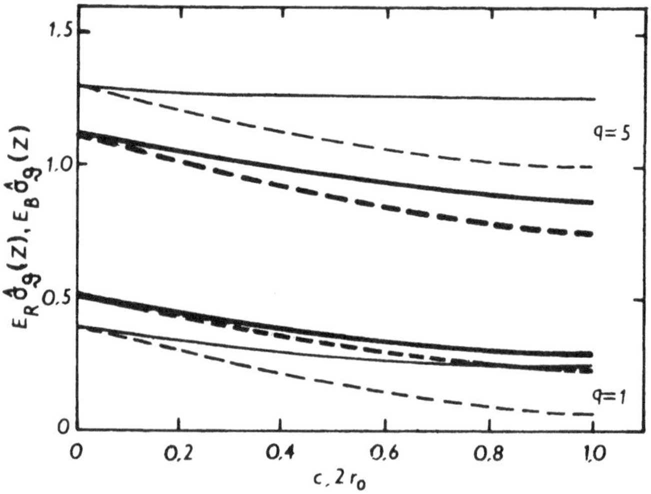

Fig. 2 The dependence of $E_R\,\hat{\sigma}_\vartheta(Z)$ (thin lines) and $E_B\,\hat{\sigma}_\vartheta(Z)$ (thick lines) on the particle size for the disc $Z = B(\sigma, r_0)$ (dotted lines) and the segment $Z = S(\sigma, c)$ (full lines) at two different values of the ratio q; a = 1.

Boolean models: As $\lambda \sim 1/q$ and $\ast \sim 1/\sqrt{q}$, then at $q \to \infty$ $E_B\,\hat{\sigma}_\vartheta \to 1/(2\sqrt{\lambda}) - \tilde{b}/(2\pi)$, i.e. again $E_B\,\hat{\sigma}_\vartheta(B) \to E_B\,\hat{\sigma}_\vartheta(\sigma) - r_0$. Similarly, $E_B\hat{\sigma}^2_\vartheta \to E_B\hat{\sigma}^2_\vartheta(\sigma) - \tilde{b}/(2\sqrt{\lambda})$, hence $E_B\hat{\sigma}^2_\vartheta(B) \to$ $\to E_B\hat{\sigma}^2_\vartheta(\sigma) - \pi a r_0 \sqrt{q}$. The asymptotic formulae for rectangles follow immediately.

Literature

[1] Coleman, R.: J. Appl. Prob. 6 (1969), 430-441.
[2] Enns, E.G., P.F.Ehlers: J. Appl. Prob. 15 (1978), 144-152.
[3] Enns, E.G., P.F.Ehlers: Austr. J. Statist. 22 (1980), 60-68.
[4] Fullman, R.L.: Trans. Met. Soc. AIME 197 (1953), 447-451.
[5] Gurland, J.: Quantitative Microscopy (DeHoff, R.T., F. N. Rhines, Eds.). McGraw-Hill Book Comp., New York 1968.
[6] Kingman, J.F.C.: J. Appl. Prob. 6 (1969), 660-672.
[7] Matheron, G.: Random Sets and Integral Geometry. J. Wiley & Sons, New York, London 1975.
[8] Santaló, L.A.: Rev. Union Mat. Argentina 9 (1943), 145-164.
[9] Santaló, L.A.: Integral Geometry and Geometric Probability. Addison-Wesley, Reading (Mass.) 1976.
[10] Saxl, I., J.Rataj: Proceedings of Colloquium on Appl. Stereology and Image Analysis, High Tatras 1988 (to be published also in Acta Stereol.)
[11] Serra, J.: Image Analysis and Mathematical Morphology. Academic Press, London 1982.
[12] Stoyan, D., W.S.Kendall, J.Mecke: Stochastic Geometry and Its Applications. Akademie-Verlag, Berlin 1987.
[13] Zähle, M.: Math. Nachr. 108 (1982), 49-72.

ON MEANS, MEDIANS, AND VARIANCES OF RANDOM COMPACT SETS

Dietrich Stoyan [1)]

1. Introduction

This paper discusses some variants of definition of mean, median, and variance of a random compact set(rcs) X. These definitions generalize corresponding definitions of these quantities for real-valued random variables or use properties of them, reformulated in terms of random sets. The author discusses four such possibilities, which use

- the indicator function of X,
- random selector points in X,
- functions describing X or the boundary of X, and
- generalizations of optimization problems which yield mean etc. in the case of real-valued random variables.

These concepts are not independent; in some cases they lead to the same characteristics.

In this paper, a rcs X is a random element of K, where K denotes the set of all compact subsets of R^d; the corresponding σ-algebra is the Borel σ-algebra with respect to the Hausdorff metric h.

It is always assumed that $0 < E(X) < \infty$, where V is the d-dimensional Lebesgue measure.

2. Definitions Using the Indicator Function

The indicator function $1_X(.)$ describes the rcs X in a very natural manner. Mean, median, and variance of $1_X(x)$, $x \in R^d$, lead to measurable deterministic functions in R^d which can be called "mean function of X" etc.. Let $p_X(x) = E1_X(x) = P(1_X(x) = 1) = P(x \in X)$.

Mean $\quad M_i(X) = \{p_X(x)\}_{x \in R^d}$

Median $\quad Med_i(X) = \{m(x)\}_{x \in R^d}$

with $m(x) = 1$ if $p_X(x) \geq 1/2$ and $=0$ otherwise

Variance $Var_i(X) = \{v(x)\}_{x \in R^d}$, $v(x) = p_X(x)(1 - p_X(x))$.

The median can be associated with a compact set, the closed hull of the set with indicator function m(x). Also this set is denoted by $Med_i(X)$.

[1)] Sektion Mathematik, Bergakademie Freiberg, Freiberg, 9200, DDR

These three characteristics are related to optimization problems. The first one is: Determine a measurable function f on R^d which minimizes

$$E(\int_{R^d} [1_X(x) - f(x)]^2 \, dx).$$

Its solution is p_X. The variance v(x) is the minimum value of $E(1_X(x) - f(x))^2$. Clearly, $v(x) \equiv 0$ if X is deterministic.
It may be also natural to define a variance of X which is a number:

$$\mathrm{var}_1(X) = E(\int_{R^d} [1_X(x) - p_X(x)]^2 \, dx).$$

The second optimization problem is: Determine a measurable function g on R^d which minimizes

$$E|g| = E(\int_{R^d} |1_X(x) - g(x)| \, dx) .$$

Its solution is m, because m(x) minimizes $E|1_X(x) - g(x)|$. The mean $E|m|$ can be rewritten in terms of set theory:

$$E|m| = EV(X \Delta \mathrm{Med}_1(X)),$$

where "Δ" is the symmetric difference operator. Clearly, $\mathrm{Med}_1(X)$ is a solution of a third problem, that of minimizing

$$EV(X \Delta C)$$

for compact subsets C of R^d. This modifies a result of Vorob'ev [10], who considered sets with a finite number of elements (e.g. points on a lattice) and suitable measures.
It is obvious that for an isotropic X the function p_X takes the same value for all x with $\|x\| = r$. Thus the notation $p_X(r)$ is used. In this case $\mathrm{Med}_1(X)$ is a ball.
Note a simple fact. If the first optimization problem is modified such that the function f has to be an indicator function of a compact set, then its solution is also m. Of course, in general it is not true

$$EV(X) = V(\mathrm{Med}_1(X)).$$

This fact suggests a further definition $M_V(x)$ of a mean of X, which was also given by Vorob'ev [10] in the case of sets with a finite number of elements. In a modified form suitable for the case of compact subsets of R^d with positive volume this definition is as follows.
Let S_p be the set $\{x \in R^d: p_X(x) \geq p\}$. Suppose that there exists a p_V with

$$V(S_{p_V}) = EV(X).$$

Then the Vorob'ev mean is defined by $M_V(X) = S_{p_V}$.
(If p_V is not unique, then the infimum of the possible values can be taken.)
The compact set $M_V(X)$ is the solution of the third optimization problem (minimization of $EV(X \Delta C)$) under the constraint $V(C) = EV(X)$.
This bases on the following theorem, which is a modification of theorem 3 in [10].

Theorem. a) Let $0 \leq p \leq 1$. Then $EV(X \Delta S_p) \leq EV(X \Delta B)$ for any Borel set B with $V(B) = V(S_p)$.

b) Let A be a Borel set with $V(A) = V(S_p)$ for some p and

$$EV(X \Delta A) \leq EV(X \Delta B)$$

for all Borel sets B with $V(A) = V(B)$. Then $V(S_p \Delta A) = 0$.

3. Definitions Using Selector Points

A selector (point) of X is a random point ξ in R^d defined over the same probability space $[\Omega, \mathcal{A}, P]$ as X with $\xi \in X$ with probability one and $E\xi \in R^d$. The idea to use selectors for defining a mean of a rcs comes from Aumann[1]:

$$M_A(X) = \{ E\xi : \xi \text{ is a selector} \}.$$

If the mean of the Hausdorff distance of X to $\{o\}$ is finite, then $M_A(X)$ is compact. It is also convex if the σ-algebra $\mathcal{A}$ is "rich enough". The latter property is a great disadvantage of this mean definition;it is not very interesting for a non-convex set, for which $M_A(X)$ is the Aumann mean of the convex hull of X.

Further properties of $M_A(X)$ are obtained if the relation to the support function description of convex sets is used, see Section 4.

Analogously, a median could be defined:

$$\text{Med}_A(X) = \{ \text{med}\ \xi : \xi \text{ is a selector} \}.$$

Here medξ is the geometric median of ξ.

As it seems, this median was not yet studied. It is easy to show by means of counterexamples that $\text{Med}_A(X)$ is not necessarily convex(also if $\mathcal{A}$ is very great).

Kruse[3] defined a variance for subsets of R^1, following the same pattern:

$$\text{Var}_A(X) = \{ \text{var}\ \xi : \xi \text{ is a selector} \}.$$

This variance has a very unpleasant property: It can be nontrivial for a deterministic X.

4. Definitions Using Functions

It is popular to describe compact sets by functions, in particular convex and starshaped ones. (A set is said to be starshaped if $x \in A$ implies $\lambda x \in A$ for all $0 \leq \lambda \leq 1$.)

A starshaped set can be described by the radius-vector function r_X,

$$r_X(u) = \sup\{ \lambda \geq 0 : \lambda u \in X \}, \quad u \in S^{d-1},$$

where S^{d-1} denotes the surface of the unit sphere of R^d.

If the rcs X is starshaped, then for every u the value $r_X(u)$ is a random variable. Its distribution function is denoted by F_u. Clearly,

$$p_X(x) = 1 - F_u(t) \quad \text{for } x = ut.$$

Let the means of $r_X(u)$ be uniformly bounded. Then the closed hull of all segments $[o, Er_X(u)\cdot u]$ is called the radius-vector mean $M_r(X)$. Analogously, median and variance can be defined. For a deterministic set X it is $Var_r(X) = \{o\}$. For any starshaped X it is $M_r(X) \subsetneqq M_A(X)$.

Convex sets are usually described by their support functions. For such a set X the support function s_X is defined by

$$s_X(u) = \max_{x \in X} (xu) \quad , \ u \in S^{d-1}.$$

This is a continuous function on S^{d-1}, and an expectation Es_X could be defined by the Bochner integral. This mean coincides with a "weak mean" which is formed for all u individually by

$$(Es_X)(u) = E(s_X(u)) \ , \ u \in S^{d-1}.$$

It is very convenient that this mean is connected with the Aumann mean by

$$s_{M_A(X)}(u) = Es_X(u), \ u \in S^{d-1},$$

for convex X; if X is non-convex, then X has to be replaced by its convex hull.

Clearly, if X is isotropic, then $M_r(X)$ and $M_A(X)$ are balls. For convex X the mean $M_A(X)$ has the following simple property: Let b(X) be the average breadth of X. Then

$$Eb(X) = b(M_A(X)).$$

Variances of convex rcs's can be defined using the variances of $s_X(u)$; this would be a "weak" variance definition. Two "strong" versions are

$$var_I(X) = \inf \{ E \| s_X - s_C \|^2 : C \in C(K) \}$$

and

$$var_E(X) = E \| s_X - s_{M_A(X)} \|^2 \quad ,$$

where s_C is the support function of C and C(K) is the set of all convex compact subsets of R^d ; $\| \cdot \|$ denotes here the sup-norm of the Banach space of all continuous functions on S^{d-1}. Unfortunately, these two variances can be different, as can be shown by examples. A general inequality is

$$var_I(X) \leqq var_E(X).$$

Näther and Albrecht[6] calculated var_E for random intervals.

Note that the minimization problem connected with var_I is the starting point of a further mean definition. Any set C which yields the infimum is called a Frechet mean $M_F(X)$ of X (with respect to the Hausdorff metric)[11]. In general, $M_A(X)$ and $M_F(X)$ do not coincide.

A median could be defined analogously.

Other possibilities of characterizing the mean behaviour of rcs's are mean-value functionals, for example the generating functional corresponding to s_X, if X is convex.

5. Characteristics for Some Stochastic Models of RCS's

For various stochastic models of rcs's it is possible to determine some of the characteristics discussed above. In the following some interesting results are presented.

Poisson polygon Π [8]

Because Π (if shifted with centre of gravity into o) is isotropic, $M_A(\Pi)$ is a circle. Its diameter is equal to the mean average breadth of Π, which is $\frac{4}{\pi}\rho^{-1}$. Consequently, the area of $M_A(\Pi)$ is $\frac{4}{\pi}\rho^{-2}$, which is the same as the mean area of Π [5]. An analogous relation for the volume of three-dimensional Poisson polyhedron is not true.

Dirichlet cell D [8]

For a typical cell (centred at o) of a Dirichlet tessellation with respect to a stationary Poisson point process of intensity μ the function $p_D(r)$ is given by

$$p_D(r) = \exp(-\mu\pi r^2), \quad r \geq 0.$$

For non-Poisson processes this probability can be expressed in terms of the Palm distribution.

Convex hull of random points in the unit circle [7]

In the unit circle n points are scattered independently, randomly and isotropically. The convex hull of these points is denoted by H_n . If the distribution of points is uniform, then $p_{H_n}(r)$ can be calculated [2]. In particular, $p_{H_n}(0) = 1 - n/2^{n-1}$. This shows that $Med_i(H_3) = \emptyset$. The diameter of $M_A(H_n)$ or the mean average breadth of H_n can be determined also in the case of a non-uniform distribution.

References

[1] Aumann, R.J.:Integrals of set-valued functions. J.Math.Anal.Appl. 12(1965),1-12.
[2] Buchta, C.: On nonnegative solutions of random systems of linear inequalities. Discrete Comput. Geom. 2(1987), 85-95.
[3] Kruse, R.: On the variance of random sets. J.Math.Anal. Appl. 122 (1987), 469-473.
[4] Matheron, G.: Random Sets and Integral Geometry. J.Wiley, 1975.
[5] Mecke, J.: Extremal properties of some geometric processes. Acta Appl. Math. 9(1987), 61-70.
[6] Näther, W. and Albrecht, M.: Linear regression with random fuzzy observations (submitted to Statistics
[7] Schneider, R.: Random approximation of convex sets. J.Microsc. (in print).
[8] Stoyan, D., Kendall, W.S. and Mecke, J.: Stochastic Geometry and Its Applications. J.Wiley, 1987.

[9] Vitale, R.: An alternate formulation of mean value for geometric figures. J.Microsc. (in print).
[10] Vorob'ev, O.Yu.: Mean-value modelling (Russ.). Nauka, Moscow 1984.
[11] Ziezold, H.: On expected figures in the plane. This volume.

ON EXPECTED FIGURES IN THE PLANE

Herbert Ziezold [1])

To K. Krickeberg on the Occasion of His 60th Birthday

In this paper a figure is an equivalence class in $(\mathbb{R}^2)^k$ modulo the group of Euclidean motions in $\mathbb{R}^2$. In the space of all figures a metric d is defined based on the Euclidean metric in $\mathbb{R}^{2k}$. If X is a random figure, an expected figure of X is an expected element μ in the sense of M. Fréchet [1], minimizing the expectation $E(d(X,\mu)^2)$. An iterative algorithm for computing mean figures of given figures $\bar{x}_1,\ldots,\bar{x}_n$ is presented.

Keywords: expected figure, mean figure, expected shape, expected element.

1. Introduction

If you are interested in the mathematical description and analysis of two-dimensional shapes of some sort of objects, you may define and measure some characteristic points of the objects with respect to any Cartesian coordinate system. You get for every object a k-ad $(x_1,\ldots,x_k)$ of points $x_i \in \mathbb{R}^2$. We say that two objects have the same figure, if their k-ads are congruent modulo translations and rotations in $\mathbb{R}^2$. If two figures are not identical, the k-ads can be put in an optimal position to each other to give the best fit with respect to the Euclidean metric in $\mathbb{R}^{2k}$. The Euclidean distance of the k-ads in an optimal position gives the metric in the space of all figures.

By Fréchet's definition of expected elements in metric spaces, [1], we can define expected figures of a random figure and mean figures of a sample of figures. Uniqueness is not guaranteed in general, but in practically relevant cases we have got it.

We cite some theoretical results proved in [4], and we develop an iterative algorithm for computing a mean figure of a sample of figures. We close with some remarks for applications.

In [3] and in a lot of papers of D. G. Kendall, e. g. [2], a shape of a k-ad is defined as the equivalence class modulo translations, rotations, and dilatations in $\mathbb{R}^2$. After norming the k-ads with respect to their centroids, see [2], page 83, the corresponding figures are representatives of the shapes and our metric for these figures is proportional to the square root of the coefficient of mismatch for the shapes in [2]. (Compare formula (21) in [2] with formula (14) below.) In this way our theory can be coupled with Kendall's theory of shape.

---

[1]) Fachbereich Mathematik, Gesamthochschule Kassel, Heinrich-Plett-Str. 40, D-3500 Kassel

2. Expected elements in metric spaces

Let X be a random element in a metric space $(\mathfrak{X},d)$. If for any element $a\in\mathfrak{X}$ the expected value $E(d(X,a)^2)$ is finite, we call every element $\mu\in\mathfrak{X}$ with

$$E(d(X,\mu)^2) = \inf_{a\in\mathfrak{X}} E(d(X,a)^2) \tag{1}$$

an expected element of X (Fréchet [1]). The set of all expected elements of X is denoted by $\mathbf{E}(X)$.

If there is an expected element μ of X, we call

$$\sigma = \sigma_X = (E(d(X,\mu)^2))^{1/2} \tag{2}$$

the standard deviation of X and σ^2 the variance of X.

Analogously, an element $m\in\mathfrak{X}$ is called **a mean of $x_1,\ldots,x_n\in\mathfrak{X}$**, if

$$\sum_{i=1}^{n} d(x_i,m)^2 = \inf_{a\in\mathfrak{X}} \sum_{i=1}^{n} d(x_i,a)^2 . \tag{3}$$

The set of all means of $x_1,\ldots,x_n$ is denoted by $\mathbf{M}(x_1,\ldots,x_n)$.

If for $n\geq 2$ there is a mean $m\in\mathbf{M}(x_1,\ldots,x_n)$, we call

$$s = \Big(\frac{1}{n}\sum_{i=1}^{n} d(x_i,m)^2\Big)^{1/2} \tag{4}$$

the standard deviation of $x_1,\ldots,x_n$ and s^2 the (empirical) variance of $x_1,\ldots,x_n$.

These definitions are generalisations of the expected value of real-valued random variables using the least-square property. But expected elements are not always uniquely defined. Furthermore the generalisation of the proposition 'The expected value of the conditional expectation of X given any sub-algebra equals the expected value of X' is not valid here. (See example 2 in [4].) But there is a strong law of large numbers in separable metric spaces:

Theorem. *Let $(\Omega,\mathfrak{F},P)$ be a probability space and let $(\mathfrak{X},d)$ be a separable metric space. Let $X_1,X_2,\ldots$ be a sequence of independent, identically distributed random elements $X_i:\Omega\to\mathfrak{X}$, with $E(d(X_i,a)^2)<\infty$ for all $a\in\mathfrak{X}$, $i=1,2,\ldots$. Then for almost all $\omega\in\Omega$*

$$\bigcap_{k=1}^{\infty} \overline{\bigcup_{n=k}^{\infty} \mathbf{M}(X_1(\omega),\ldots,X_n(\omega))} \subseteq \mathbf{E}(X_1), \tag{5}$$

where $\overline{D}$ denotes the closure of a set D. ([4])

Using the ideas of the last part of the proof of this theorem one gets the

Corollary. *Suppose $X_1,X_2,\ldots$ are as in the Theorem. Let $S_n(\omega)$ be the standard deviation of $X_1(\omega),\ldots,X_n(\omega)$, $\omega\in\Omega$, $n\in\mathbf{N}$, and let σ be the standard deviation of the X_i. If $(\mathfrak{X},d)$ is compact and separable, then for almost all $\omega\in\Omega$*

$$\lim_{n\to\infty} S_n(\omega) = \sigma. \tag{6}$$

3. The metric space of figures in the plane

For the description of the following computations it is more convenient to describe the points in the plane by complex numbers instead of pairs of real numbers. For $x=(x_1,\dots,x_k)\in\mathbf{C}^k$ we set

$$\|x\| = \Big(\frac{1}{k}\sum_{i=1}^{k} x_i\overline{x_i}\Big)^{1/2} \tag{7}$$

(We propose the factor $1/k$ in order to be able to compare the distances of figures of given real objects for different k.)

Let $\mathbf{S}$ be the unit circle $\{x\in\mathbf{C} : |x|=1\}$ and define $1=(1,\dots 1)\in\mathbf{R}^k$. For $u\in\mathbf{S}$, $a\in\mathbf{C}$ and $x\in\mathbf{C}^k$ ux is the rotation of the k-ad x by u and $x+a1$ is the translation of the k-ad x by a.

We define a distance δ in $\mathbf{C}^k$ by

$$\delta(x,y) = \inf_{\substack{u\in\mathbf{S}\\ a\in\mathbf{C}}} \|x-uy-a1\| \quad \text{for } x,y\in\mathbf{C}^k. \tag{8}$$

For every $x\in\mathbf{C}^k$ we define $\tilde{x}$ as the equivalence class of x modulo δ and call it **the figure given by** x. Thus figures are equivalence classes of k-ads of points in $\mathbf{C}$ modulo Euclidean motions in $\mathbf{C}$ for some fixed k. The space of all figures with its quotient metric is the metric space $(\mathfrak{X},d)$ we are considering now.

For determining mean figures in the sense of the preceding section we need the following definition.

Definition. *We say that k-ads $x_i=(x_1^{(i)},\dots,x_k^{(i)})\in\mathbf{C}^k$, $i=1,2,\dots,n$, are **in optimal position to each other**, if*

$$\sum_{1\le i<j\le n}\delta(x_i,x_j)^2 = \inf_{\substack{u_1,\dots,u_n\in\mathbf{S}\\ a_1,\dots,a_n\in\mathbf{C}}} \sum_{1\le i<j\le n}\delta(u_ix_i+a_i1,u_jx_j+a_j1)^2 \tag{9}$$

Proposition 1 in [4] assures that k-ads $x_1,\dots,x_n$ are in optimal position to each other if and only if

the centroids $\frac{1}{k}\sum_{l=1}^{k} x_l^{(i)}$ are all equal,

$m=(m_1,\dots,m_k)$ with $m_l=\frac{1}{n}\sum_{i=1}^{n} x_l^{(i)}$ defines a mean of $x_1,\dots,x_n$, and

for every $i=1,\dots,n$ the k-ad x_i is in optimal position to m.

Furthermore we get in this case for the variance of $\tilde{x}_1,\dots,\tilde{x}_n$

$$s_n^2 = \frac{1}{n^2}\sum_{1\le i<j\le n}\delta(x_i,x_j)^2. \tag{10}$$

This proposition has the implication that we can consider only centered k-ads. If c is the centroid of $\mathbf{x}\in\mathbf{C}^k$, we call $\mathbf{x}^* = \mathbf{x} - c\mathbf{1}$ the centered k-ad to $\mathbf{x}$.

Another implication is that $\mathbf{M}(\tilde{\mathbf{x}}_1, \ldots, \tilde{\mathbf{x}}_n)$ is not empty for all $\mathbf{x}_1, \ldots, \mathbf{x}_n \in \mathbf{C}^k$. This follows by the compactness of $\mathbf{S}$.

A generalisation of this result for random figures is given in [4], proposition 2.

Let us now investigate how to compute the optimal position of two k-ads $\mathbf{x} = (x_1, \ldots, x_k)$ and $\mathbf{y} = (y_1, \ldots, y_k) \in \mathbf{C}^k$ to each other. Let $\langle \mathbf{x}, \mathbf{y} \rangle$ be the Hermitian form underlying the norm $\|\cdot\|$, i. e.

$$\langle \mathbf{x}, \mathbf{y} \rangle = \frac{1}{k} \sum_{j=1}^{k} x_j \overline{y_j} \ . \tag{11}$$

We have to find $v\in\mathbf{S}$ with

$$\| \mathbf{x}^* - v\mathbf{y}^* \| = \inf_{u\in\mathbf{S}} \| \mathbf{x}^* - u\mathbf{y}^* \| \ , \tag{12}$$

i. e. we have to minimize

$$\| \mathbf{x}^* - u\mathbf{y}^* \|^2 = \| \mathbf{x}^* \|^2 + \| \mathbf{y}^* \|^2 - (\langle \mathbf{x}^*, \mathbf{y}^* \rangle \overline{u} + \langle \mathbf{y}^*, \mathbf{x}^* \rangle u) . \tag{13}$$

If $\langle \mathbf{x}^*, \mathbf{y}^* \rangle = Rv$ with $R\geq 0$ and $v=\exp(i\varphi)\in\mathbf{S}$, then (13) is minimized by $u=v$ and maximized by $u=-v$. Thus we get the

Proposition. *Given* $\mathbf{x},\mathbf{y}\in\mathbf{C}^k$, let $Rv = \langle \mathbf{x}^*, \mathbf{y}^* \rangle$ *with* $R\geq 0$ *and* $v\in\mathbf{S}$. *Then*

$$\delta(\mathbf{x},\mathbf{y})^2 = \| \mathbf{x}^* - v\mathbf{y}^* \|^2 = \| \mathbf{x}^* \|^2 + \| \mathbf{y}^* \|^2 - 2R \tag{14}$$

and

$$\sup_{u\in\mathbf{C}} \| \mathbf{x}^* - u\mathbf{y}^* \|^2 - \inf_{u\in\mathbf{C}} \| \mathbf{x}^* - u\mathbf{y}^* \|^2 = \| \mathbf{x}^* + v\mathbf{y}^* \|^2 - \| \mathbf{x}^* - v\mathbf{y}^* \|^2 = 4R. \tag{15}$$

Corollary 1. *The k-ads $\mathbf{x}^*, \mathbf{y}^* \in\mathbf{C}^k$ are in optimal position to each other if and only if $\langle \mathbf{x}^*, \mathbf{y}^* \rangle \in [0,\infty[$.*

Corollary 2. *If $R>0$, there is only one mean figure of $\tilde{\mathbf{x}}$ and $\tilde{\mathbf{y}}$.*

In case of $R=0$ the k-ads $\mathbf{x}^*$ and $u\mathbf{y}^*$ are in optimal position to each other for every $u\in\mathbf{S}$ and there are infinitely many mean figures of $\tilde{\mathbf{x}}$ and $\tilde{\mathbf{y}}$.

4. An algorithm for computing mean figures to more than two figures

Example 2 at the end of [4] shows that we cannot compute mean figures of $\tilde{x}_1, \tilde{x}_2, \tilde{x}_3$ by generalizing the above considerations to weighted means of two figures and applying this to the computations of $\tilde{x}' \in M(\tilde{x}_1, \tilde{x}_2)$ and of $\tilde{x}'' \in M(\tilde{x}', \tilde{x}', \tilde{x}_3)$ to get a mean figure of $\tilde{x}_1, \tilde{x}_2, \tilde{x}_3$.

Instead we proceed in the following way to get a minimum for the the function $y \to f(y) = \sum_{i=1}^{n} \delta(y, x_i)^2$ for given $x_1, \ldots, x_n \in \mathbf{C}^k$:

We first center the given k-ads and we get $x_1^*, \ldots, x_n^*$. To get a sequence $\tilde{y}_1, \tilde{y}_2, \ldots$ which converges to a mean figure $\tilde{y}'$ of $\tilde{x}_1, \ldots, \tilde{x}_n$, we choose any k-ad y_0, e. g. $y_0 = x_1^*$, and we iterate the following steps to get $y_1, y_2, \ldots$ and new positions $x_i^{(j)}$, $x_i^{(0)} = x_i^*$, i=1,...,n, for j=1,2,...:

Step 1. For every i=1,2,...,n compute $u_i^{(j)} \in \mathbf{S}$ such that $x_i^{(j+1)} = u_i^{(j)} x_i^{(j)}$ and y_j are in optimal position to each other, i. e. $u_i^{(j)} = \langle y_j, x_i^{(j)} \rangle / |\langle y_j, x_i^{(j)} \rangle|$, if the denominator is not zero, otherwise set $u_i^{(j)} = 1$.

Step 2. Compute $_{+1} = \frac{1}{n} \sum_{i=1}^{n} x_i^{(j+1)}$.

We set $c_j = (\frac{1}{n} \sum_{i=1}^{n} \| y_j - x_i^{(j)} \|^2)^{1/2}$. If $\tilde{y}_j \neq \tilde{y}_{j+1}$, we get

$$c_j \geq (\frac{1}{n} \sum_{i=1}^{n} \| y_j - x_i^{(j+1)} \|^2)^{1/2} > (\frac{1}{n} \sum_{i=1}^{n} \| y_{j+1} - x_i^{(j+1)} \|^2)^{1/2} = c_{j+1}.$$

Thus the sequence $c_1, c_1, \ldots$ converges to a number c'. The sequence $y_1, y_2, \ldots$ contains a subsequence $y_{j_1}, y_{j_2}, \ldots$ converging to a centered k-ad y'. If the sample $x_1, \ldots, x_n$ is an independent realisation of a diffuse distribution in $\mathbf{C}^k$, f has almost surely got a local minimum at y'. To get a global minimum with some certainty you may repeat this iteration procedure with different initial k-ads, e. g. you may choose y_0 from $\{x_1^*, \ldots, x_n^*\}$ at random. If the smallest value for c' results several times, that might be the global minimum and the limits $\tilde{y}'$ are mean figures in this case.

We have not yet got general criteria to see whether there are unique limits c' and $\tilde{y}'$. Such criteria may be something like

> 'The variations of the component points of the figures *between* the figures may not be to great relative to the distances of the points *within* the figures'.

5. Practical Experiences. Hints for applications

Our experience with the above algorithm is that in practically relevant problems of determining mean figures of k-ads $x_1, \dots, x_n$ of real objects we have got the global minimum c' and a unique $\tilde{y}'$ for every initial k-ad y_0.

To explore the range of uniqueness of c' and $\tilde{y}'$ we generated k-ads $x_1, \dots, x_n$ by choosing the j-th point of x_i for fixed $\alpha \in [0, 2\pi]$ at random from the set $A_j^{(k)}(\alpha) = \{x \in C : |x| \le 1, |\frac{2\pi j}{k} - \arg(x)| \le \frac{\alpha}{2}\}$, $j=1,2,\dots,k$; $i=1,2,\dots,n$. For $\alpha=\pi$, $k=3$, and $n=3,4,5,10$ we repeated this process 500 times, and in each case we got maximally 3 samples $(x_1, \dots, x_n)$ with more than one local minimum for f. We believe that this demonstrates an acceptable reliability of the proposed algorithm for determining mean figures, as in these cases the component points may lie very chaotically in the plane.

For applications in the taxonomy of several species we propose to compute the mean and the variance of the figures of the sample objects for every species. Then one can compare the variance of the weighted mean figures, which is defined by an obvious generalisation of the definition in Section 2, with the weighted mean of the variances of the figures for the different species, as it is done in the F-statistics, to see whether there is an obvious difference between the figures of the species. Furthermore one can classify a new object to one of the species by the distances of its figure to the mean figures of the species.

Until now we have not yet got distributional statements. But our theory may though be useful for exploratory purposes.

6. References.

1. M. FRÉCHET: Les éléments aléatoires de nature quelconque dans un éspace distancié. *Ann. Inst. H. Poincaré* **10** (1948), 215-310.
2. D. G. KENDALL: Shape manifolds, procrustean metrics, and complex projective spaces. *Bull. London Math. Society* **16** (1984), 81-121.
3. D. STOYAN, W. S. KENDALL, and J. MECKE: Stochastic Geometry and Its Applications. *Akademie-Verlag, Berlin*, 1987. *Wiley, Chichester*, 1987.
4. H. ZIEZOLD: On expected figures and a strong law of large numbers for random elements in quasi-metric spaces. *Transactions of the Seventh Prague Conference on Information Theory, Statistical Decision Functions, Random Processes and of the Eighth European Meeting of Statisticians (Tech. Univ. Prague, Prague, 1974), Vol. A*, 591-602. *Reidel, Dordrecht*, 1977.

THE JOHNSON-MEHL TESSELLATION WITH TIME DEPENDENT NUCLEATION INTENSITY IN VIEW OF BASIC 3-D TESSELLATIONS

Vratislav Horálek [1)]

1. Introduction

Let us assume that three-dimensional Euclidean space E^3 is divided into spacefilling and non-overlapping three-dimensional domains C_i, bounded and connected closed subsets of E^3, forming so called cells. Their boundaries ∂C_i are piecewise smooth closed surfaces separating C_i from its exterior. Let $C_* = \{C_i\}$ be the set of all cells in E^3, then $\mathcal{C} \subset C_*$ is said to be a 3-d tessellation. An extensive review of papers on 3-d tessellations can be found e.g. in Stoyan's et al book [8] and Møller's booklet [7].

In Section 2 we recall the model assumptions of two basic types of random tessellations : the Poisson-Voronoi tessellation (the PV-tessellation) derived from the original Voronoi model [9] and the time-homogeneous Johnson-Mehl tessellation (the THJM-tessellation) suggested by Johnson and Mehl [4] for the description of the reaction kinetics in processes of nucleation and growth. In Section 3 we investigate the properties of the time-nonhomogeneous Johnson-Mehl tessellation (the TNHJM-tessellation) with nucleation intensity

$$\alpha(t) = \alpha t^{\beta} , \qquad (1)$$

$\beta > -1$ and $\alpha > 0$ being constants and t time. In Section 4 the corresponding 'ergodic' moments of this tessellation are compared with those of the PV-tessellation and the THJM-tessellation. The found internal relationships between TNHJM-tessellation of quoted type and the above mentioned 3-d tessellations extend the results derived by Miles [6].

2. Characterization of PV, THJM and TNHJM tessellations

Let (x, y, z) be coordinates of E^3 space. The PV-tessellation is created under the following assumptions:

(a) the point patterns representing the centres of gravity $P_i(x, y, z)$ of nuclei have fixed positions in E^3 and form the homogeneous Poisson point process with intensity α in E^3;

(b) the cell growth starts in all points P_i at the same instant, the growth rate v being constant and the growth isotropic;

(c) the cell C_i growing from the point P_i contains all points $P \in E^3$ closer to P_i than to any P_j, $i \neq j$, i. e.

---

1) National Research Institute for Machine Design,
190 11 Praha 9 - Běchovice II, Czechoslovakia

$$C_i = \{ P \in E^3 : \| P - P_i \| \leq \| P - P_j \| \ , \ i \neq j \} \ ,$$

where $\|$ denotes the Euclidean distance. Therefore, in the PV-tessellation generated from points $\{P_i\}$ the cells C_i are spacefilling convex polyhedra.

In the THJM-tessellation the formation of cells C_i is more complicated. The physical background of this dynamic model was formulated by Johnson and Mehl [4]. From the mathematical point of view their assumptions can be transformed in this way:

(a) the cell growth does not start in all points P_i at the same instant as in the PV-tessellation, but the starting instants t_i , $i = 1, 2, \ldots,$ are different;

(b) the points (P_i, t_i) , $i = 1, 2, \ldots,$ form the homogeneous Poisson point process with intensity α in E^4 space with coordinates (x, y, z, t), $t \geq 0$, i.e. the probability that a nucleus is activated in a hitherto untransformed volume element dV in the time interval $(t, t + dt)$ is equal to $\alpha \, dV \, dt$;

(c) the real activation of a nucleus is associated with only those points (P_i, t_i) which are located in hitherto untransformed volume at the instant t_i ; the nucleation intensity α per unit time is constant;

(d) the cell C_i grows from the point P_i after its activation, the growth rate v being constant and the growth isotropic;

(e) the cell C_i contains all points $P \in E^3$ which such a property that the sphere growing from the point (P_i, t_i) by a constant rate v is the first one to reach P i.e.

$$C_i = \{ P \in E^3 : r_j - r_i > v(t_i - t_j) \ ; \ i \neq j \} \ ,$$

$r_k = \| P - P_k \|$, $k = i, j$, being the Euclidean distance of P_k to P. Therefore in the THJM-tessellation the cells are non-convex and, in general, they are 'star-shaped' with respect to their nuclei.

From the point of view of the activation of nuclei, the above mentioned THJM model can be taken as a model characterized by a homogeneous Poisson process with intensity α , this process being a homogeneous immigration process with a constant spatial nucleation intensity. A generalization of this approach can be provided in such a sense that the homogeneous immigration process is replaced by a nonhomogeneous one with time dependent intensity of immigration - see e.g. Horálek [2].
For such a generalized model Miles [6] derived the 'ergodic' moments of the volume V , the surface area S , the total edge length L_1 and the number of vertices N_0 of the 'typical cell' in the following forms

$$E(V) = \left\{ \int_0^\infty J(t) \ \alpha(t) \ dt \right\}^{-1} \ , \qquad (2)$$

$$E(S) = \frac{64\pi^2 v^5}{3} \frac{\int_0^\infty J(t)\, I^2(t)\, dt}{\int_0^\infty J(t)\, \alpha(t)\, dt}, \tag{3}$$

$$E(L_1) = \frac{64\pi^4 v^7}{5} \frac{\int_0^\infty J(t)\, I^3(t)\, dt}{\int_0^\infty J(t)\, \alpha(t)\, dt}, \tag{4}$$

$$E(N_o) = \frac{1024\pi^5 v^9}{105} \frac{\int_0^\infty J(t)\, I^4(t)\, dt}{\int_0^\infty J(t)\, \alpha(t)\, dt}, \tag{5}$$

where

$$I(t) = \int_0^t (t - x)\, \alpha(x)\, dx \tag{6}$$

and

$$J(t) = \exp\left\{-\frac{4}{3}\pi v^3 \int_0^t (t - x)^3\, \alpha(x)\, dx\right\}. \tag{7}$$

3. The TNHJM-tessellation with nucleation intensity (1)

Inserting (1) into (2) to (7), all integrals involved there convert into gamma functions $\Gamma\{n(\beta)\}$, where the argument depends on the exponent β in (1) and the 'ergodic' moments $E(V)$ to $E(N_o)$ become the functions of parameters β, v and α. Hence for the TNHJM-tessellation with intensity (1) we have [3]

$$E(V) = k_V(\beta) \left\{ \frac{v^{3(\beta+1)}}{\alpha^3} \right\}^{\frac{1}{\beta+4}}, \tag{8}$$

$$E(S) = k_S(\beta)\, N_V^{-\frac{2}{3}}; \quad E(L_1) = k_{L1}(\beta)\, N_V^{-\frac{1}{3}}; \quad E(N_o) = k_{NO}(\beta), \tag{9}$$

where $N_V = 1/E(V)$ is the mean number of cells per unit volume and

$$k_V(\beta) = \frac{1}{W}\left\{2^{3(\beta+1)}\pi^{\beta+1}(\beta+4)^3 Z^{-(\beta+1)}\right\}^{\frac{1}{\beta+4}}, \tag{10}$$

$$k_S(\beta) = \frac{8}{3}(\beta+4)\left(\frac{\pi}{WZ}\right)^{\frac{1}{3}}\Gamma\left(\frac{2\beta+7}{\beta+4}\right), \tag{11}$$

$$k_{L1}(\beta) = \frac{4}{5}(\beta+4)^2\pi\left(\frac{\pi}{WZ}\right)^{\frac{2}{3}}\Gamma\left(\frac{3\beta+10}{\beta+4}\right), \tag{12}$$

$$k_{NO}(\beta) = \frac{32}{105}(\beta+4)^3\frac{\pi^2}{WZ}\Gamma\left(\frac{4\beta+13}{\beta+4}\right) \tag{13}$$

and

$$W = \left(\frac{\beta + 1}{\beta + 4}\right) \quad ; \qquad Z = (\beta + 1)(\beta + 2)(\beta + 3) \quad .$$

The knowledge of coefficients $k_V(\beta)$, $k_S(\beta)$ and $k_{L1}(\beta)$ enables us to determine the coefficients of further descriptors of the spatial structure and of the induced structure observable in a random plane sectioning the cell structure. For the spatial structure we shall consider the following descriptors : S_V, the average surface area per unit volume, and $E(H)$, the mean length of cell orthogonal projection onto an isotropic random line; for the induced structure they shall be : $E(A)$, the mean value of cell area section; N_A, the mean number of cell sections per unit area; N_L, the mean number of cells intercepted by a line of unit length and finally $E(L)$, the mean value of the intercept length through a 'typical cell'. We have

$$S_V = k_{SV}(\beta)\, N_V^{\frac{1}{3}} = \frac{1}{2}\, k_S(\beta)\, N_V^{\frac{1}{3}} , \tag{14}$$

$$E(H) > k_H(\beta)\, N_V^{-\frac{1}{3}} = \frac{1}{12}\, k_{L1}(\beta)\, N_V^{-\frac{1}{3}} , \tag{15}$$

$$E(A) < k_A(\beta)\, N_V^{-\frac{2}{3}} = \frac{12}{k_{L1}(\beta)}\, N_V^{-\frac{2}{3}} , \tag{16}$$

$$N_A > k_{NA}(\beta)\, N_V^{\frac{2}{3}} = \frac{1}{k_A(\beta)}\, N_V^{\frac{2}{3}} , \tag{17}$$

$$N_L = k_{NL}(\beta)\, N_V^{\frac{1}{3}} = \frac{1}{k_L(\beta)}\, N_V^{\frac{1}{3}} , \tag{18}$$

$$E(L) = k_L(\beta)\, N_V^{-\frac{1}{3}} = \frac{4}{k_S(\beta)}\, N_V^{-\frac{1}{3}} . \tag{19}$$

The numerical values of the coefficients $k_i(\beta)$, $i = V, S, L_1$ and N_0 are introduced in Tab. 1 for chosen values of $\beta \in \langle -1; \infty)$. By using of them the coefficients $k_i(\beta)$, $i = S_V, H, A, N_A, N_L$ and L can be easy calculated from (14) to (19).

4. Properties of coefficients $k_i(\beta)$

Now, we investigate the behaviour of coefficients $k_i(\beta)$ and compare the morphometric characteristics of individual tessellations under consideration.

As it follows from Tab. 1 and from the expressions (10) to (19) the coefficients $k_i(\beta)$, $i = S, L_1, N_o, H, S_V$ are decreasing functions and coefficients $k_i(\beta)$, $i = A, L$, are increasing functions in the interval $\langle -1; \infty)$ and all have finite positive limits for β tending to -1 and to ∞. But the coefficient $k_V(\beta)$ is a unimodal function with maximum 1.2647 at the point $\beta = 0.56$ and it converges to 0 when β tends to -1 and to ∞.

Tab. 1 Coefficients $k_i(\beta)$, $i = V, S, L_1, N_0$

β	$k_V(\beta)$	$k_S(\beta)$	$k_{L1}(\beta)$	$k_{N0}(\beta)$
-1.0	0	5.821	17.496	27.071
-0.8	0.245	5.623	16.628	25.567
-0.6	0.518	5.467	15.977	24.493
-0.4	0.768	5.340	15.467	23.687
-0.2	0.970	5.233	15.055	23.060
0	1.116	5.143	14.714	22.559
0.2	1.209	5.065	14.428	22.149
0.4	1.255	4.997	14.182	21.807
0.6	1.264	4.936	13.968	21.518
0.8	1.245	4.883	13.781	21.270
1.0	1.206	4.835	13.615	21.055
2.0	0.894	4.653	13.006	20.303
∞	0	3.906	10.782	18.047

Of course, for $\beta = 0$ the TNHJM model with intensity (1) must convert into the THJM model with constant intensity α . Really, all coefficients in the row $\beta = 0$ of Tab. 1 agree with those gained by Meijering [5] , Gilbert [1] and Miles [6] for the THJM-tessellation.

Now, let us investigate the limits of coefficients (10) to (13) for β tending to -1 . In this case, we have to state that they converge to the values relevant to the PV-tessellation (see the three papers mentioned above), but due to the requirement that β must be larger than -1 , they never can reach them; note that for the PV-tessellation it holds $E(V) = 1/\alpha$ and the term in brackets in (8) converges to $1/\alpha$ as β tends to -1 . Taking into account the different physical conditions for creation of the PV-tessellation and of the TNHJM-tessellation, presented in Section 2, this is very surprising result. It shows a significant influence of nucleation intensity $\alpha(t)$ on the morphometric characteristics of real cell structure and this fact can play a very important role in modelling material structures with requested technical properties by using the TNHJM-tessellation that seems as more flexible than the THJM-tessellation.

Literature

[1] Gilbert, E. N.: Random subdivisions of space into crystals. Ann. Math. Statist. 33 (1962) 3, 958-972.

[2] Horálek, V.: On some types of non-homogeneous birth-immigration-death processes. Aplikace matematiky ČSAV 9 (1964) 6, 421-434.

[3] Horálek, V.: A note on the time-nonhomogeneous Johnson-Mehl tessellation. Advances in Appl. Probab. 20 (1988) 3, 684-685.

[4] Johnson, W. A., Mehl, R.: Reaction kinetics in processes of nucleation and growth. Trans. of AIME 35 (1939), 416-458.

[5] Meijering, J. L.: Interface area, edge length and number of vertices in crystal aggregates with random nucleation. Philips Res. Report 8 (1953) 4, 270-290.

[6] Miles, R. E.: The random division of space. Proc. of the Symp. on Statist. and Probab. Problems in Metallurgy. Adv. in Appl. Probability, Special Suppl., 1972, 243-266.

[7] Møller, J.: Random tessellations in R^d. Memoirs No. 9, Dept. of Theor. Statist., Inst. of Math., University of Aarhus, 1986.

[8] Stoyan, D., Kendall, W. S., Mecke, J.: Stochastic Geometry and Its Applications. Akademie-Verlag, Berlin, 1987.

[9] Voronoi, G.: Nouvelles applications des parameters continus a la théorie des formes quadratiques. J. Reine Angew. Math. 134 (1908), 198-287

MATHEMATICAL ASPECTS OF THE STEREOLOGICAL MEMBRANE PROBLEM

J. Mecke [1)]

Abstract

The stereological problem of determining a membrane thickness distribution from orthogonal intercepts was treated by Cruz-Orive, Gundersen, Jensen, Østerby and others. This paper contains some additional remarks. The existence of probability densities for the thickness distribution is no longer required. An other stochastic geometric model is presented which leads to the same integral equation as the membrane problem. It suggests a family of thickness distributions which are well adapted to the problem.

1. Introduction

Cruz-Orive [2], Gundersen, Jensen and Østerby [3] have shown how to reconstruct the distribution of membrane thickness from that of orthogonal intercept length, measured in isotropic uniform random planes (IUR-planes). The present paper contains some additional remarks.

In section 2, the problem will be formulated in an abstract mathematical manner. A general solution is presented, where the existence of a probability density for the distribution of membrane thickness is not required. The notations are partly adopted from article [3].

Section 3 deals with a simple stochastic geometric model leading to the same integral equation as the membrane problem. This model suggests distributions of membrane thickness which together with the corresponding distributions of orthogonal intercepts are closely related to well-known elementary distributions. These distributions, which are fitted to the problem, are specified in section 4.

Statistical applications of the results are suggested but not implemented.

Coleman [1] and Cruz-Orive [2] also considered the case of intercepts from IUR-lines. For this model similar results may be derived as above, see the remark at the end of section 3.

1) Friedrich-Schiller-Universität Jena, Sektion Mathematik, UHH, 17. OG., DDR-6900 Jena

2. Problem and solution

We consider a random arrangement Φ of planes in the 3-dimensional Euclidean space with the property that the distribution of Φ is invariant under all proper motions; Φ is said to be an isotropic stationary plane process.

Let E be a fixed plane. The probability density function p of the typical angle Θ between E and a random plane from Φ is given by

$$p(x) = 4\pi^{-1}\sin^2 x; \qquad 0 \leq x \leq \pi/2. \tag{2.1}$$

Let ξ be a plane parallel to E at a random distance T, where T is independent of Φ. Then E and ξ may be considered the boundaries of a membrane, from which strips are cut out by the random planes of Φ. The width of the typical strip is denoted by L. The random variables L, T, Θ are related by the equation

$$1/L = (1/T)\sin\Theta , \tag{2.2}$$

where T and Θ are independent, cf. p. 178 in [2] and formula (2) in [3].

The distribution function of T is denoted by G, that of 1/T by G_1. The distribution of L always has a density f. The probability density f_1 of 1/L fulfils $f_1(x)=x^{-2}f(1/x)$. From (2.2) it can be easily deduced that

$$f_1(x) = 4\pi^{-1}x^2 \int_{(x,\infty)} t^{-2}(t^2-x^2)^{-1/2}\, dG_1(t). \tag{2.3}$$

If G_1 has a density g_1, then (2.3) turns into equation (5) in [3] or (1), (2) in [2]. The function g_1 can be reestablished from f_1 by solving the integral equation of Abel type (2.3), a solution is presented by formula (15) in [3]:

$$g_1(x) = 2^{-1}x^3 \int_x^\infty \left[2f_1(t)-tf_1'(t)\right] t^{-3}(t^2-x^2)^{-1/2}\, dt.$$

A general inversion formula not assuming the existence of a density for G_1 is given by

$$1-G_1(x+) = 2^{-1}\int_x^\infty f_1(t)\left[t(t^2-x^2)^{-1/2} + t^{-1}(t^2-x^2)^{1/2}\right]dt. \tag{2.4}$$

Here $G_1(x+)$ means the limit on the right of G_1 at x, i.e. $G_1(x+) = \mathrm{prob}(1/T \leq x)$. Formula (2.4) can be proved by substituting (2.3) into (2.4): Then the right-hand side of (2.4) changes into the expression

$$2\pi^{-1} \int_{(x,\infty)} s^{-2} \int_x^s t^2(s^2-t^2)^{-1/2}\left[t(t^2-x^2)^{-1/2} + t^{-1}(t^2-x^2)^{1/2}\right] dt dG_1(s).$$

Finally, the substitution

$$z = \arcsin\left[(s^2-t^2)^{1/2}(s^2-x^2)^{-1/2}\right]$$

transforms the inner integral into

$$\int_0^{\pi/2} \left[s^2+(s^2-x^2)\cos(2z)\right] dz = \pi s^2/2.$$

Define a function $u: (0,\infty)^2 \longrightarrow [0,\infty)$ by

$$u(t,x) = \begin{cases} 2^{-1}\left[t(t^2-x^2)^{-1/2}+t^{-1}(t^2-x^2)^{1/2}\right], & \text{if } t > x, \\ 0, & \text{if } t \leq x. \end{cases}$$

Then (2.4) implies

$$1 - G_1(x+) = \mathbb{E}\, u(1/L,x), \tag{2.5}$$

where $\mathbb{E}$ is the symbol for mathematical expectation. Hence, if $l_1,\ldots,l_n$ are observations of L, then

$$n^{-1}\left[u(1/l_1,x) + \ldots + u(1/l_n,x)\right]$$

represents a reasonable unbiased estimate of $1-G_1(x+)$.

Since $1-G_1(x+)=G(1/x)$, (2.5) is equivalent to the following equation for the distribution function G of T:

$$G(x) = \mathbb{E}\, u(1/L, 1/x). \tag{2.5'}$$

3. Equivalent problem

Let $\underline{U}=(U_1,U_2,U_3,U_4)$ be an isotropic random vector in $\mathbb{R}^4$, i.e. a random 4-dimensional vector the components of which are elements of the real axis $\mathbb{R}$, and the distribution of which is invariant under all rotations of $\mathbb{R}^4$ about the origin. Its length will be denoted by $|\underline{U}|$ and the length of its projection $Q\underline{U}=(U_1,U_2,U_3)$ onto the (x_1,x_2,x_3)-subspace by $|Q\underline{U}|$:

$$|\underline{U}|= (U_1^2 + U_2^2 + U_3^2 + U_4^2)^{1/2}, \quad |Q\underline{U}| = (U_1^2 + U_2^2 + U_3^2)^{1/2}.$$

Then we have

$$|Q\underline{U}| = |\underline{U}| \sin \Gamma, \tag{3.1}$$

where Γ is the angle between $\underline{U}$ and the x_4-axis; $0 \leq \Gamma \leq \pi/2$. The random variables $|\underline{U}|$ and Γ are independent.

The probability density of $\sin\Gamma$ is equal to that of $X^{1/2}(X+Y)^{-1/2}$, where X is a random variable having a χ^2-distribution with 3 degrees of freedom, and Y is a random variable independent of X having a χ^2-distribution with 1 degree of freedom. This can be deduced from (3.1) specifying the distribution of $\underline{U}$ to be a rotation invariant 4-dimensional normal distribution. This means that $\sin^2\Gamma$ has a beta distribution of first kind with the density

$$b(x) = 2\pi^{-1}x^{1/2}(1-x)^{-1/2}$$

on (0,1). It follows that the angle Γ has the probability density p given by (2.1), i.e. Γ and Θ are identically distributed.

Hence, equation (2.2) corresponds to (3.1) by the substitution $1/L \rightarrow |Q\underline{U}|$, $1/T \rightarrow |\underline{U}|$, $\Theta \rightarrow \Gamma$. In this way, it becomes clear that equations (2.3), (2.4) are valid for $|\underline{U}|$, $|Q\underline{U}|$ if the distribution function of $|\underline{U}|$ is denoted by G_1 and the probability density of $|Q\underline{U}|$ by f_1.

Remarks

- To the stereological membrane problem, where the section is made by a line instead of a plane, there corresponds the above vector model with projection onto the (x_1,x_2)-plane instead of the (x_1,x_2,x_3)-subspace.
- If a d-dimensional sheet in $\mathbb{R}^d$ is intersected by an isotropic stationary arrangement of s-dimensional flats (s-flat process, $1 \leq s \leq d-1$), then there is a correspondence in the above sense to the projection of an isotropic random vector in $\mathbb{R}^{d+1}$ onto an (s+1)-dimensional subspace.

4. Special distributions

Let the components U_1, U_2, U_3, U_4 of $\underline{U}$ be independent and identically normally distributed with mean 0 and variance $(2b)^{-1}$. Then $|\underline{U}|^2$ has a gamma distribution with parameters 2,b, i.e. with the density

$$g(2,b;x) = b^2\, x \exp(-bx) \tag{4.1}$$

and $|Q\underline{U}|^2$ has a gamma distribution with parameters 3/2, b, i.e. with the density

$$g(3/2,b;x) = 2b^{3/2}\pi^{-1/2}x^{1/2}\exp(-bx) \tag{4.2}$$

on $(0,\infty)$. In other words, $2b|\underline{U}|^2$ is χ^2-distributed with 4 degrees of freedom and $2b|Q\underline{U}|^2$ is χ^2-distributed with 3 degrees of freedom. The distribution of $|\underline{U}|^2$ is equal to the convolution of an exponential distribution with itself (2-stage Erlang distribution with mean 2/b). The distribution of $|Q\underline{U}|$ is called a Maxwell distribution.

Taking into account the correspondence to the membrane problem and the linearity of (2.3), we realize:

Proposition 4.1. If $1/T^2$ is distributed according to the density

$$q_1 g(2,b_1;\cdot) + \dots + q_n g(2,b_n;\cdot), \tag{4.3}$$

then $1/L^2$ is distributed according to the density

$$q_1 g(3/2,b_1;\cdot) + \dots + q_n g(3/2,b_n;\cdot), \tag{4.4}$$

where $n=1,2,\dots$; $q_1,\dots,q_n \in \mathbb{R}$, $q_1+\dots+q_n=1$; $b_1,\dots,b_n > 0$, and $g(2,b;\cdot)$, $g(3/2,b;\cdot)$ are given by (4.1), (4.2).

In case all weights $q_1,\dots,q_n$ are nonnegative, the linear combinations (4.3), (4.4) are said to be mixings of gamma densities. There exist indeed examples of nonnegative densities (4.3), (4.4), where some of the coefficients $q_1,\dots,q_n$ are negative, e.g. $n=2$, $b_1=1$, $b_2=2$, $q_1=5/4$, $q_2=-1/4$.

Perhaps proposition 4.1 may contribute to a parametric approach to the statistical problem of reconstructing the distribution of 1/T from observations of 1/L. Already a linear combination (4.3) with n=2 has 3 free parameters.

The distribution (4.3) is not as special as it seems for the present. The following proposition means that any distribution on $(0,\infty)$ can be approximated by a distribution of type (4.3).

Proposition 4.2. Let μ be any probability measure on $(0,\infty)$. Then there exists a sequence of probability measures $\mu_1, \mu_2, \dots$ on $(0,\infty)$ with densities of type (4.3), where not all coefficients are necessarily nonnegative, which weakly converges towards μ.

P r o o f . Let $X, Y_1, Y_2, \dots$ be independent, positive random variables, where X has the distribution μ, and $Y_1, Y_2, \dots$ are identically distributed according to an exponential density

$$e(b;x) = b\exp(-bx);\ x > 0,$$

with b=1. The law of large numbers implies that the random variables

$$Z_n = n^{-1}(2Y_1/1 + 3Y_2/2 + \dots + (n+1)Y_n/n)$$

tend in probability to 1. Consequently, the distributions of XZ_n converge weakly to μ.

Using Laplace transforms and decomposition into partial fractions, we find that Z_n is distributed according to a density

$$a_{n1}e(n/2;\cdot) + a_{n2}e(2n/3;\cdot) + \dots + a_{nn}e(n^2/(n+1);\cdot)$$

with not necessarily nonnegative coefficients $a_{n1},\dots,a_{nn}$. It follows that the distribution μ_n of XZ_n has a density of the form

$$\int e(b;\cdot)\alpha_n(db), \tag{4.5}$$

where α_n is a signed measure on $(0,\infty)$. The density (4.5) is equal to

$$\int_0^\infty b^{-2}\left(\int_0^b t\,\alpha_n(dt)\right)g(2,b;\cdot)db, \tag{4.5'}$$

which may be looked on as a continuous linear combination of gamma densities. But (4.5') can be approximated in the weak sense with any accuracy by discrete linear combinations (4.3).

References

[1] Coleman, R.: An introduction to mathematical stereology. Memoirs no. 3, Math. Dept., University of Aarhus, 1979.

[2] Cruz-Orive, L.M.: Estimation of sheet thickness distributions from linear and plane sections. Gegenbaurs morph. Jahrb. 126 (1980). 175-188.

[3] Jensen, E.B., Gundersen, H.J.G., Østerby, R.: Determination of membrane thickness distribution from orthogonal intercepts. Journal of Microscopy, Vol. 115 (1979), Pt 1, 19-33.

STEREOLOGICAL ESTIMATION OF SECOND- AND HIGHER-ORDER PROPERTIES OF RANDOM SETS

E.B. Jensen[1], K. Kiêu[2] and H.J.G. Gundersen[3]

1. Introduction and summary

Second- and higher-order properties of random sets can be described by reduced moment measures. In the present paper, we will present stereological estimators of reduced moment measures of random d-dimensional sets in $\mathbb{R}^n$, $d \leq n$, of a very general type: The class of random d-Hausdorff rectifiable closed sets, cf. Zähle (1982). The reduced moment measures can be expressed as moments with respect to a random set which is isotropic but not stationary. The geometric measure decomposition presented in Jensen and Gundersen (1988) is just the one needed for treating such random sets. It will be shown that the q'th reduced moment measure can be estimated from observations on a q-dimensional section.

A more detailed version of the results of the present paper can be found in Jensen et al. (1988). Similar results have been obtained independently by Schwandtke (1988) and Zähle (1988). Estimation of non-reduced moment measures has been discussed in Nagel (1987).

2. Random sets and moment measures

Let λ_n^d be the d-dimensional Hausdorff measure in $\mathbb{R}^n$, $d=0,1,\ldots,n$. In particular, $\lambda_n^n = \lambda_n$ is the n-dimensional Lebesgue measure. Let N_d be the set of closed subsets $\varphi \subseteq \mathbb{R}^n$ which are λ_n^d-rectifiable. Denote by $\mathcal{B}^n$ the Borel σ-algebra in $\mathbb{R}^n$. A random (d,n)-set Φ is then a random variable taking values in N_d. We will in what follows assume that Φ is stationary and isotropic.

The q'th moment measure is the measure on $\mathcal{B}^{qn}$ defined by

$$\Lambda_\Phi^{(q)}(B) = E \prod_{i=1}^{q} \lambda_n^d(\Phi \cap B_i)$$

when $B = B_1 \times \ldots \times B_q$, $B_i \in \mathcal{B}^n$, $i=1,\ldots,q$. Note that $\Lambda_\Phi^{(1)}(B) = E\, \lambda_n^d(\Phi \cap B) = \nu_n^d\, \lambda_n(B)$, because of the stationarity of Φ. ν_n^d is called the intensity of Φ.

The moment measures can be expressed in terms of the reduced moment measures. These are defined via the Palm distribution. The Palm distri-

---

1: Institute of Mathematics, Aarhus University, Ny Munkegade, DK-8000 Aarhus C
2: Institut National de la Recherche Agronomique, Jouy-en-Josas
3: Stereological Research Laboratory, Aarhus University

bution P_0 of a random (d,n)-set Φ can be interpreted as its conditional distribution given that $0 \in \Phi$. P_0 is uniquely defined by the property

$$E \int_\Phi h(x,\Phi)\lambda_n^d(dx) = \nu_n^d \int_{\mathbb{R}^n} E_0(h(x,\Phi+x))\lambda_n(dx) ,$$

where E_0 is the mean in the Palm distribution. If we let Φ_0 be a random (d,n)-set with distribution P_0, the reduced moment measures are then defined by

$$K_\Phi^{(q)}(B) = \Lambda_{\Phi_0}^{(q-1)}(B)/(\nu_n^d)^{q-1}, \quad B \in \mathcal{B}^{(q-1)n} .$$

Moment measures can be expressed in terms of reduced moment measures. For instance, when $q = 2$, we have

$$\Lambda_\Phi^{(2)}(B_1 \times B_2) = (\nu_n^d)^2 \int_{B_1} \int_{\mathbb{R}^n} I_{B_2}(x+h) K_\Phi^{(2)}(dh)\lambda_n(dx) .$$

3. Stereological estimation of reduced moment measures

Below we give a stereological estimator of $\Lambda_{\Phi_0}^{(q)}(B_1\times\ldots\times B_q)$ for rotation invariant B_i's. The estimator involves measurement on and in the neighbourhood of a fixed q-dimensional affine subspace.

Let $\mathcal{L}_{q(0)}$ be the set of q-dimensional linear subspaces $L_{q(0)}$ of $\mathbb{R}^n$. Let $dL_{q(0)}$ be the element of the rotation invariant measure on $\mathcal{L}_{q(0)}$. $dL_{q(0)}$ is uniquely defined when it is scaled such that the total integral equals $c(n,q)$, a well-known constant (cf. e.g. Jensen et al., 1988).

The estimator is based on a generalized version of Blaschke-Petkantschin formula. Define the mapping

$$(x_1^d,\ldots,x_q^d) \in \varphi^d \to (L_{q(0)}, x_1^{d-n+q},\ldots,x_q^{d-n+q})$$

where $L_{q(0)}$ is the q-subspace containing $x_1^d,\ldots,x_q^d$ and x_i^{d-n+q} is the position of x_i^d inside $\varphi \cap L_{q(0)}$, $i=1,\ldots,q$. From the coarea formula we know that there exists a Jacobian $J_n^{d,q}$ such that

$$J_n^{d,q}(x_1^d,\ldots,x_q^d;\varphi)dx_1^d \ldots dx_q^d = dx_1^{d-n+q}\ldots dx_q^{d-n+q}\, dL_{q(0)}$$

($dx_i^d = \lambda_n^d(dx_i)$). Now we can state the first step towards the development of a stereological estimator.

Lemma 1. Let $L_{q(00)}$ be a fixed q-subspace and let

$$f(\varphi; L_{q(00)}, B_1\times\ldots\times B_q) =$$

$$c(n,q) \int_{\varphi\cap B_1\cap L_{q(00)}} \ldots \int_{\varphi\cap B_q\cap L_{q(00)}} J_n^{d;q}(x_1^{d-n+q},\ldots,x_q^{d-n+q};\varphi)^{-1} dx_1^{d-n+q}\ldots dx_q^{d-n+q}$$

Then

$$(\nu_n^d)^q K_\Phi^{(q+1)}(B_1\times\ldots\times B_q) = E_0\, f(\Phi; L_{q(00)}, B_1\times\ldots\times B_q),$$

$$q = \max(1,n-d),\ldots,n.$$

Proof.

$$(\nu_n^d)^q K_\Phi^{(q+1)}(B_1\times\ldots\times B_q)$$

$$= E_0 \int_{\Phi\cap B_1}\ldots\int_{\Phi\cap B_q} dx_1^d\ldots dx_q^d$$

$$= \int_{L_{q(0)}} E_0 \int_{\Phi\cap B_1\cap L_{q(0)}}\ldots\int_{\Phi\cap B_q\cap L_{q(0)}} (J_n^{d;q})^{-1} dx_1^{d-n+q}\ldots dx_q^{d-n+q} dL_{q(0)}$$

The isotropy of Φ implies the isotropy of P_0 which in turn implies that the integrand of the outer integral does not vary with $L_{q(0)}$. The lemma follows. □

It may be of practical interest that when $q \geq n-d+1$ the qth reduced moment measure can be expressed as the mean in P_0 of a measurement along a random $q-1$ flat moving in a fixed $L_{q(00)}$.

Next lemma gives a procedure for selecting typical points in $\Phi\cap L_{q(00)}$. Let $L_{r(00)}$ be a fixed r-subspace in $L_{q(00)}$, $\max(1,n-d) \leq r \leq q$. Let $G_n^{d,r}(\cdot;\varphi)$ be the Jacobian of the mapping

$$x^d \in \varphi \rightarrow (x^{n-r}, x^{d-n+r}),$$

where x^{n-r} is the projection of x^d onto $L_{r(00)}^\perp$ and x^{d-n+q} represents the position of x^d on $\varphi\cap L_{r(00)}+x^{n-r})$.

Lemma 2. Let T_r be a bounded Borel subset of $L_{r(00)}$ and let h be a measurable non-negative function, $h\colon N_d \rightarrow \mathbb{R}_+\cup\{0\}$ Then,

$$E\int_{\Phi\cap T_r} h(\Phi-x)\,dx^{d-n+r} = \nu_n^d\, \lambda_r(T_r)\, E_0(h(\Phi)\, G_n^{d;r}(0;\Phi))$$

Proof. Let I_{n-r} be the unit cube in $L^{\perp}_{r(00)}$. Applying the Palm distribution definition, we easily get

$$\nu^d_n \lambda_n(T_r \oplus I_{n-r}) E_0(h(\Phi) G^{d;r}_n(0;\Phi))$$

$$= E \int_{\Phi \cap (T_r \oplus I_{n-r})} h(\Phi - x^d) G^{d;r}_n(0;\Phi - x^d)\, dx^d$$

$$= \int_{I_{n-r}} E \int_{\Phi \cap (T_r + x^{n-r})} h(\Phi - x^{n-r} - x^{d-n+r})\, dx^{d-n+r} dx^{n-r}$$

By the stationarity of Φ, the integrand of the outer integral does not vary with x^{n-r}. The lemma follows. □

We can now state the theorem which gives an unbiased estimator of the reduced moment measures. The theorem easily follows from the two lemmas.

Theorem. Let $L_{q(00)}$ and $L_{r(00)}$ be fixed q- and r-subspaces such that $L_{r(00)} \subseteq L_{q(00)}$ and let $T_r \subseteq L_{r(00)}$ be a bounded Borel subset of $L_{r(00)}$. Then

$$(\nu^d_n)^q K^{(q+1)}(B_1 \times \ldots \times B_q)$$

$$= \frac{1}{\nu^d_n \lambda_r(T_r)} E \int_{\Phi \cap T_r} f(\Phi - x; L_{q(00)}, B_1 \times \ldots \times B_q) / G^{d;r}_n(0;\Phi - x)\, dx^{d-n+r}$$

where $\max(1, n-d) \le r \le q$. □

Special cases of the theorem are estimators of the reduced second moment measure of planar fibre processes, cf. Stoyan (1981), and spatial surface processes, respectively.

4. Discussion

The use of moment measures in the analysis of spatial structure is not common except in the case $d = 0$ (point processes), probably because of a total randomness reference such as the Poisson point process is in general missing. Total randomness is defined for some random processes of closed sets such as line processes or random processes of closed sets which can be described by marked point processes. In the latter model class, total randomness is given by the Boolean model. The reduced second moment measure of the Boolean model can be expressed in terms of the intensity ν^d_n and a functional of the typical grain distribution. In Jensen et al. (1988) a modified reduced second moment measure is suggested. This alternative moment measure can be estimated by the same stereological procedures as standard reduced moment measures and allows a straightforward comparison with the Boolean model of a random process which can be described by a marked point process.

References

Jensen, E.B. and Gundersen, H.J.G. (1988): Fundamental stereological formulae based on isotropically oriented probes through fixed points - with applications to particle analysis. To appear in Proceedings of the 4th International Conference in Stereology and Stochastic Geometry (special issue of Journal of Microscopy).

Jensen, E.B., Kiêu, K. and Gundersen, H.J.G. (1988): On stereological estimation of reduced moment measures. Research Report No. 174, Department of Theoretical Statistics, Institute of Mathematics, University of Aarhus.

Nagel, W. (1987): An application of Crofton's formula to moment measures of random curvature measures. Forschungsergebnisse N/87/12, Friedrich-Schiller-Universität, Sektion Mathematik, Jena.

Schwandtke, A. (1988): Second-order quantities for stationary weighted fibre processes. Math. Nachr. 139, to appear.

Stoyan, D. (1981): On the second-order analysis of stationary planar fibre processes. Math. Nachr. 102, 189-199.

Zähle, M. (1982): Random processes of Hausdorff rectifiable closed sets. Math. Nachr. 108, 49-72.

Zähle, M. (1988): A kinematic formula and moment measures of random sets. To appear.

SOME REMARKS ON STEREOLOGICAL ESTIMATION OF PARTICLE DENSITIES

Annette Schwandtke[1)]

1. Introduction

The aim of this paper is to present a strong proof that two estimators of particle density are unbiased: the estimator of N_A using the Gundersen frame and the disector estimator of N_V. Further, it is shown that in some cases it is useful to estimate the specific connectivity number rather than the particle density. It is assumed that the particle systems can be described by stationary marked point processes [5]. This paper continues the work in [3].

2. Two Particle Number Estimators

For the determination of the planar particle density N_A Gundersen [1] suggested the following estimator:

$$\hat{N}_A = N^F/A(W). \tag{1}$$

It is assumed that the particle system is observed in a rectangular window W of area A(W). The quantity N^F is the number of particles which intersect W but do not hit the thick line in Fig. 1.
(Note that for the determination of N^F sometimes information from outside of W is necessary.)

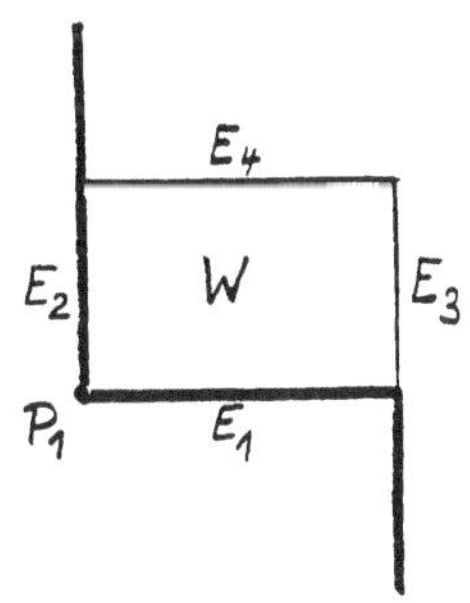

Fig. 1. The Gundersen frame

1) Annette Schwandtke Bergakademie Freiberg DDR-9200 Freiberg
Bernhard-von-Cotta-Str. 2

Sterio [4] suggested an estimator of N_V, the spatial particle density, which uses the so-called disector. Here the three-dimensional particle system is intersected by two parallel planes: the counting plane e_c (taken as the (x,y)-plane) and the testing plane e_t (taken as the plane z=d). It is assumed that d is smaller than the smallest particle diameter. Consequently, no particle lies between the planes e_c and e_t without intersecting one of them. In e_c the Gundersen frame is given, see Fig. 2.

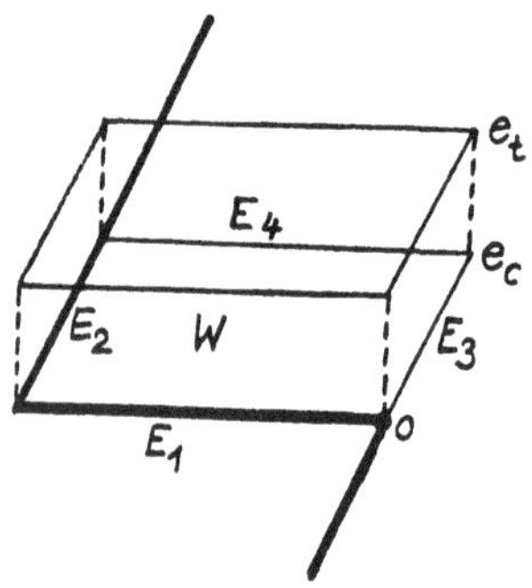

Fig. 2. The disector

A particle X is counted, if and only if

1) X intersects the window W in the counting plane e_c but does not hit the forbidden line of the frame, and
2) X does not intersect the test plane e_t.

For checking condition 2 the following basic requirement is necessary: For all particle fragments visible in the two planes of intersection it is possible to decide which particle fragments are belonging to the same particle.

Let N^D be the number of particles which satisfy these two conditions. Then N_V is estimated by

$$\widehat{N}_V = N^D/(A(W)\cdot d). \qquad (2)$$

Theorem. The estimators $\widehat{N}_A$ and $\widehat{N}_V$ are unbiased, if the corresponding particle systems can be described by a stationary marked point process.

Proof. 1) Consider a planar particle system given by the stationary marked point process $\Phi_A = \{[t_n, X_n]\}$, where $\{X_n\}$, n = 1,2,..., is a sequence of planar random compact sets. The nth particle is then the set X_n+t_n and N_A is the intensity of Φ_A.

Denote by $T(X_n)$ the set of all points $t = (x,y)$ for which a particle X_n+t is counted by the frame, see Fig. 3. Then

$$N^F = \sum_{[t_n, X_n] \in \Phi_A} 1_{T(X_n)}(t_n),$$

where $1_T(t)$ is the indicator function of T. The Campbell theorem [5] yields

$$EN^F = N_A \iint 1_{T(X)}(t)\, dt\, \mathcal{M}(dX), \tag{3}$$

where $\mathcal{M}$ denotes the mark distribution of Φ_A.

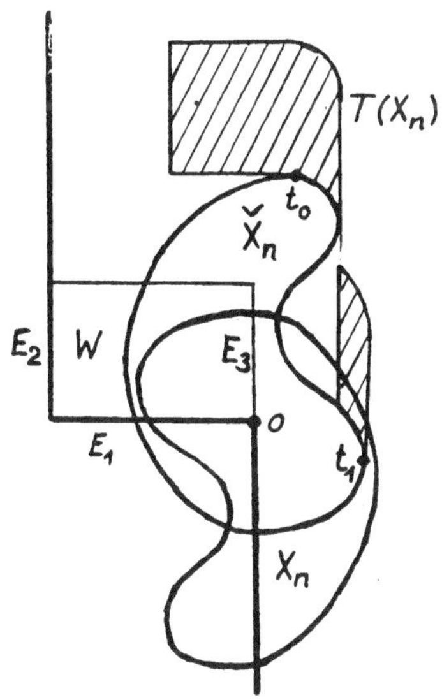

Fig. 3. Determination of the set $T(X_n)$

Remember that $T(X_n)$ is the set of points t for which X_n+t intersects W, but does not intersect the forbidden edge E_1 and the infinite continuations of the edges E_2 and E_3. Denote by $\check{X}_n$ the set $-X_n$. Let $t_o = (x_o, y_o)$ be the "y-maximal" point of $\check{X}_n$, e.g. that point with the maximal y-coordinate, and if there are two or more such points then that with the greatest x-coordinate is chosen. $t_1 = (x_1, y_1)$ is the "x-maximal" point of $\check{X}_n$, which is defined analogously. The set $T(X_n)$ is given by

$$T(X_n) = \left\{ (x,y) \in R^2 \;\middle|\; \begin{array}{l} x_1 - l(E_1) \leq x \leq x_o,\ y_o \leq y \leq y_o + l(E_2) \\ x_o < x \leq x_1,\ y_{max}(x) < y \leq y_{max}(x) + l(E_2) \end{array} \right\}$$

where $y_{max}(x)$ denotes the maximal y-coordinate corresponding to a given x-coordinate such that $(x, y_{max}(x)) \in \check{X}_n$ and $l(E_i)$ denotes the length of the edge E_i, $i = 2,3$. The area of the set $T(X_n)$ is equal to the area $A(W)$ of the frame, independently of the shape of X_n. Hence, (3) implies

$$EN^F = N_A \cdot A(W),$$

e.g., the estimator given in (1) is unbiased.

2) The unbiasedness of $\widehat{N}_V$ is proved as follows. The spatial particle system is described by a stationary marked point process $\Phi_V = \{[u_n, X_n]\}$ of intensity N_V. Now $\{X_n\}$, $n = 1,2,\ldots$, is a sequence of spatial random compact sets and u_n is the "z-maximal" point of the nth particle. Let the disector be positioned at the origin o, as shown in Fig. 2. Denote now by $U(X_n)$ the set of all points $u = (x,y,z)$ for which the particle $X_n + u_n$ is counted by the disector. Clearly, only points u with z-coordinate between 0 and d belong to $U(X_n)$. For a fixed z, $0 \leq z \leq d$, the intersection of the particle $X_n + u$, $u = (x,y,z)$, with the plane e_c is a (planar) particle $X_n' + (x,y)$. Denote by $T(X_n')$ the set of all points (x,y) for which the particle $X_n' + (x,y)$ is counted by the frame and the corresponding particle $X_n + u$ is counted by the disector. Thus,

$$N^D = \sum_{[u_n, X_n] \in \Phi_V} 1_{[0,d]}(z_n)\ 1_{T(X_n')}(x_n, y_n).$$

By the Campbell theorem

$$EN^D = N_V \int \iiint 1_{[0,d]}(z)\ 1_{T(X')}(x,y)\ dx\ dy\ dz\ \mathcal{M}(dX),$$

where $\mathcal{M}$ is the mark distribution of Φ_V.

As one has already seen in 1), the area of $T(X')$ is equal to $A(W)$ independently of the shape of X'. Hence,

$$EN^D = N_V \cdot A(W) \cdot d,$$

that means, $\widehat{N}_V$ is unbiased.

The estimator $\widehat{N}_V$ can be rewritten as

$$\widehat{N}_V = \widehat{N}_A / d,$$

where $\widehat{N}_A = N^D / A(W)$ is the unbiased Gundersen frame estimator of the specific number N_A of that particles which hit the plane e_c but do not intersect the plane e_t. Obviously, instead of $\widehat{N}_A$ any other unbiased estimator of N_A may be applied.

3. Estimators of Specific Connectivity Numbers

The estimation methods mentioned above sometimes need information from the outside of W. If the samples can be moved (under a microscope) and if the particles are visible also outside of W, then there are no problems. But for samples given on photos on which particles hitting the boundary are partially visible only, it is necessary to have an estimator of N_A using only the information inside the window.

The paper [3] proposes such a counting rule, which is similar to the Gundersen frame. This counting rule is also natural for systems Ξ of non-convex particles, because it estimates the specific connectivity (or Euler-Poincaré) number E_A [5]. For convex particles it is $E_A = N_A$. If the particle system Ξ is motion-invariant and if all particles are members of the convex ring [5], then

$$\hat{E}_A = N^{\chi}/A(W) \tag{4}$$

is an unbiased estimator of E_A. Here

$$N^{\chi} = \chi(\Xi \cap W) - \chi(\Xi \cap E_1) - \chi(\Xi \cap E_2) + \chi(\Xi \cap P_1),$$

and $\chi(K)$ denotes the connectivity number of the set K.

The unbiasedness of $\hat{E}_A$ can be easily shown using a special case of Steiner's formula for generalized Minkowski functionals [3]. Note that $\hat{E}_A$ is also unbiased, if Ξ is stationary only (not necessarily isotropic) and the particles are convex ones. This can be proved as follows: In the case of convex particles $\chi(\Xi \cap W)$, $\chi(\Xi \cap E_1)$, $\chi(\Xi \cap E_2)$ and $\chi(\Xi \cap P_1)$ are equal to the number of particles hitting W, E_1, E_2 and P_1, respectively. Clearly, N^{χ} can be written as

$$N^{\chi} = \sum_{[t_n, X_n] \in \Phi_A} (1_{W \oplus \check{X}_n}(t_n) - 1_{E_1 \oplus \check{X}_n}(t_n) - 1_{E_2 \oplus \check{X}_n}(t_n) + 1_{P_1 \oplus \check{X}_n}(t_n)),$$

where $\Phi_A = \{[t_n, X_n]\}$ is the marked point process of intensity E_A which describes the particle system Ξ and $\oplus$ denotes Minkowski addition. The Campbell theorem yields

$$EN^{\chi} = E_A \iint (1_{W \oplus \check{X}}(t) - 1_{E_1 \oplus \check{X}}(t) - 1_{E_2 \oplus \check{X}}(t) + 1_{P_1 \oplus \check{X}}(t))\, dt\, \mathcal{M}(dX)$$

$$= E_A \int (A(W \oplus \check{X}) - A(E_1 \oplus \check{X}) - A(E_2 \oplus \check{X}) + A(P_1 \oplus \check{X}))\, \mathcal{M}(dX).$$

It follows

$$EN^{\chi} = E_A \cdot A(W),$$

because for any convex set K

$$A(W \oplus K) - A(E_1 \oplus K) - A(E_2 \oplus K) + A(P_1 \oplus K) = A(W). \qquad (5)$$

Note that (5) is not necessarily true for non-convex sets.
For example, if K is an annulus,

$$K = \left\{ (x,y) \mid 0 < r_1 \leq \sqrt{x^2+y^2} \leq r_2 \right\},$$

a simple counter-example is obtained.

It can be concluded that the spatial specific connectivity number E_V can be estimated in a similar way, using the information of a spatial sample inside the window of observation only [3].

Literature

[1] Gundersen, H.G.J.: Estimators of the number of objects per area unbiased per edge effects. Microsc. Acta 81 (1978), 107-117.

[2] Gundersen, H.G.J.: Stereology of arbitrary particles. J. Microsc. 138 (1986), 3-45.

[3] Schwandtke, A., D. Stoyan, and V. Schmidt: Some remarks on the stereological estimation of particle characteristics. (in preparation)

[4] Sterio, D.G.: The unbiased estimation of numbers and sizes of arbitrary particles using the disector. J. Microsc. 134 (1984), 127-136.

[5] Stoyan, D., W.S. Kendall, and J. Mecke: Stochastic geometry and its applications. J. Wiley/Akademie-Verlag, Chichester/Berlin 1986.

ON ANISOTROPY OF SHAPE AND SPATIAL DISTRIBUTION

Viktor Beneš [1)]

1. Introduction

One of the aims of quantitative metallography is to define and evaluate suitable structural parameters in order to express the connection between the technology of material production and the geometry of the structure. In the present paper the technology in question is the extrusion of alloys which is typified by a single dominant orientation of elongated phase particles. However, the extruded structure cannot be modelled simply by means of the pressing of an isotropic structure [9] because the particles are not only pressed but also broken up and special spatial configurations such as clusters or chains appear. An empirical investigation of a plane section parallel to the dominant direction is presented, this structure is considered as a realisation of a planar germ-grain model [10] where all assumptions for stationarity are fulfilled except that the typical grain need not be independent of the point process of germs (grain centroids). The main aim of this paper is to quantify the anisotropy of the structure using a small number of parameters which become items of a database record of the alloy's properties. The problem is reduced to the evaluation of four parameters characterizing the anisotropy of (a) grain shape and orientation, (b) spatial distribution of germs, the correlation between these anisotropies and the total inner anisotropy. The first two parameters are well-known, some aspects of parametric method of their quantification are discussed. The remaining parameters are new. The application of the theoretical background of [8] allows the expression of interesting inner properties of the structure.

2. The parametric methods

Consider the stationary fibre process Ψ_1 of grain boundaries. The anisotropy of the grain shape and orientation is determined by the rose of directions. Its probability density (if it exists) $f(\beta)$ is connected with the rose of intersections $P_L(\beta)$ by the equation [4]

$$\frac{d^2}{d\beta^2} P_L(\beta) + P_L(\beta) = 2L_A f(\beta), \qquad \beta \in \langle 0,\pi) \tag{1}$$

---

1) Research Institute for Metals, Panenské Břežany, 25070, Czechoslovakia

L_A being the intensity of Ψ_1. From the many methods of estimation the one selected for discussion is that of taking several testing orientations

$$\beta_i = (2i-1)\Delta \ , \quad \Delta = \pi/2n, \quad i=1,\dots,n \tag{2}$$

with the subsequent fitting of $P_L(\beta)$ by a parametric model. Furthermore, consider the second-order analysis [7] of the stationary point process ϕ_1 (with intensity λ_1) of germs. Taking other methods [3] into account as well as the fact that for a more detailed description of the spatial distribution of points, moment measures of higher order [2] are necessary, the method is based on the reduced second order moment measure K of an anullus sector $S_{q\delta}^{\beta\Delta}$ (see Fig. 1). Its unbiased estimation in a bounded window W [6] is used for the evaluation of the anisotropic spatial distribution. The measure $K(S_{q\delta}^{\beta\Delta})$

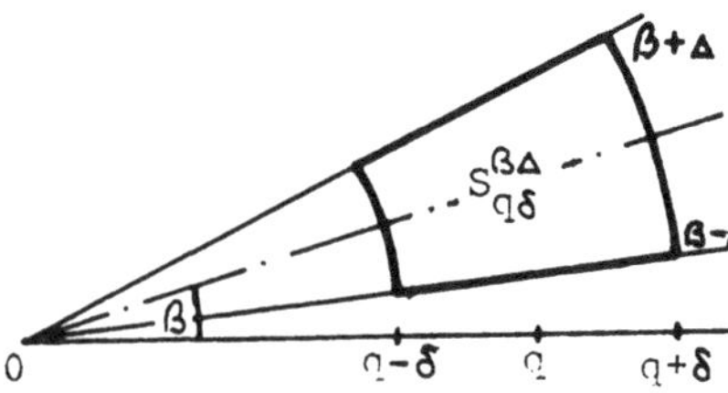

Fig. 1 - The anullus sector $S_{q\delta}^{\beta\Delta}$

depends both on Δ, δ and also on the scale factor q. To obtain a global information the whole sector $S_{qq}^{\beta\Delta}$ is used; the upper bound

$$q_o = 2q \tag{3}$$

for a square window is put equal to its side.

Assume now that the dominant direction of the structure coincides with $\beta=0$. The multiple correlation coefficient [5] between $(\cos 2\beta_i)$ and $(P_L(\beta_i),\ K(S_{q\delta}^{\beta_i\Delta}))$ was successfully applied to quantify the total anisotropy with possible decomposition to partial components. But more natural is to measure anisotropy directly rather than as a degree-of-fit to a given model. Consider a linear regression model

$$E[P_L(\beta_i)] = a \cos 2\beta_i + b, \quad \operatorname{var} P_L(\beta_i) = G \tag{4}$$

(similarly for $E[K(S_{q\delta}^{\beta\Delta})]$). In a matrix form $EY=XA$, $\operatorname{var}Y=G$, $A=\binom{a}{b}$, $X_i=(\cos 2\beta_i, 1)$, G is the covariance matrix of the correlated observation. Then the parameters are estimated by the least squares method $A=CX'G^{-1}Y$, where $C=(X'G^{-1}X)^{-1}$ has diagonal elements c_{11}, c_{22}, X' is the transpose of X. The parameter $D=|a|b^{-1}$ yields the degree of anisotropy equal to zero in the isotropic case. It can exceed 1 only in the case of extreme orientations when the model (4) does not fit well. By putting (4) into (1) and integrating over β a stereological estimator $L_A=b\pi/2$ is obtained and its variance estimator is $\operatorname{var}L_A=\pi^2 c_{22}/4$.

The Poisson cluster model is a first approximation of the extrusion structure, clusters being interpreted as broken particles. Let ϕ_1 be a Neymann-Scott process [10] such that the parent points (with intensity λ) are centroids of an ellipse $\mathcal{E}(u,v)$ with axis length u,v, and the orientation of u coincides with $\beta=0$. A Poisson random number (with parameter $\varkappa$) of daughter points are uniformly randomly scattered in each ellipse, $\lambda_1 = \lambda \varkappa$. An anisotropic version of formula 5.3.4 in [10] for the second moment measure

$$K(S_{q\delta}^{\beta\Delta}) = 4q\delta\Delta + \lambda^{-1}\left[F_w\left(\frac{q+\delta}{u},\beta+\Delta\right) - F_w\left(\frac{q+\delta}{u},\beta-\Delta\right) - F_w\left(\frac{q-\delta}{u},\beta+\Delta\right) + F_w\left(\frac{q-\delta}{u},\beta-\Delta\right)\right] \quad (5)$$

contains $w = \frac{v}{u}$ and a distribution function $F_w(q,\beta)$ of the simultaneous distribution of the distance and orientation between daughter points in a cluster ellipse $\mathcal{E}(1,w)$, which can be estimated by simulation. The best choice of q_o in (3) is $q_o=2u$ because this is the smallest sector radius containing all information on anisotropy. Consider further elliptical grains with axis s,t and orientation β_o. Using the definition of the modified support function $r(\beta)$ [10] and the equality $P_L(\beta)=2r(\beta)$, the rose of intersections of the germ-grain model (with possible overlappings) is

$$P_L(\beta) = 2\,\lambda_1 (s^2 \sin^2(\beta - \beta_o) + t^2 \cos^2(\beta - \beta_o))^{1/2}. \quad (6)$$

Parameters of functions (5), (6) can be randomized.

Example 1: In a Monte-Carlo study 500 realisations of the above germ-grain model were simulated in a square window of area 1 with $\lambda=10$, $\varkappa=15$, u=0.4, v=0.04, s=0.02, t=0.004, q=0.2, $\delta=0.2$, n=6. The matrix G was estimated for $P_L(\beta_i)$-case I, and for $K(S_{qq}^{\beta_i\Delta})$-case II, using the plus-sampling:

$$\text{I}\begin{pmatrix} 0.33 & 0.29 & 0.42 & 0.34 & 0.29 & 0.11 \\ 0.29 & 1.01 & 0.95 & 0.74 & 0.60 & 0.23 \\ 0.42 & 0.95 & 1.97 & 1.15 & 0.97 & 0.35 \\ 0.34 & 0.74 & 1.15 & 1.55 & 0.93 & 0.35 \\ 0.29 & 0.60 & 0.97 & 0.93 & 1.11 & 0.28 \\ 0.11 & 0.23 & 0.35 & 0.35 & 0.28 & 0.29 \end{pmatrix} \quad \text{II}\begin{pmatrix} 6.22 & 2.67 & 1.12 & 0.90 & 1.51 & 2.63 \\ 2.67 & 5.70 & 3.58 & 2.60 & 2.26 & 1.72 \\ 1.12 & 3.58 & 4.82 & 3.77 & 2.67 & 1.45 \\ 0.90 & 2.60 & 3.77 & 4.48 & 3.47 & 1.59 \\ 1.51 & 2.26 & 2.67 & 3.47 & 5.84 & 3.57 \\ 2.63 & 1.72 & 1.45 & 1.59 & 3.57 & 6.87 \end{pmatrix}$$

A subsequent simulation yields data in Table 1 with results of estimation.

Table 1: A numeric example 1, β_i as in (2)

i	1	2	3	4	5	6	D	L_A	$(\mathrm{var}L_A)^{1/2}$
$P_L(\beta_i)$	1.46	4.65	6.26	5.56	4.92	2.27	0.518	5.99	0.698
$K(S_{qq}^{\beta_i\Delta})$	8.25	5.79	4.95	4.91	5.75	10.82	0.243		

In the practical work of the metallographical laboratory the matrix G can be estimated from the results of the subsequent measurement of the extrusion specimens. Using G a discussion on the choice of β_i and n can be made by means of the statistical principle components method [5]. Another idea concerning the handling of the correlated data from image analysis was presented (for a different problem) in [11].

3. The inner anisotropy

Let $(R^2, \mathcal{B}^2)$ be the real plane with borel σ-algebra, $\mathcal{L} = \langle 0,\pi)$ and $\mathcal{T}$ the borel σ-algebra on $\mathcal{L}$. Suppose that $L \in \mathcal{T}$ is fixed. It is reasonable for the purposes of this paper to model the structure of extrusion plane section as a stationary marked point process Ψ [10] with the point process ϕ_1 (with intensity λ) of germs $x_i \in R^2$ and marks $m(x_i)$ which denote the length of the relevant grain boundary which has the tangent direction in L. For any bounded set $S \in \mathcal{B}^2$ we define

$$K_L(S) = \frac{1}{\lambda} E \sum_{(x,m(x))\in\Psi} 1_{\langle 0,1\rangle^2}(x) \sum_{\substack{x\neq y\in S_x \\ (y,m(y))\in\Psi}} m(y) \tag{7}$$

where $S_x = \{y \in R^2,\ y=z+x,\ z\in S\}$ for $x \in R^2$. From the refined Campbell theorem [8] follows immediately

$$K_L(S) = \iint \sum_{\substack{(y,l(y))\in\Psi \\ 0\neq y\in S}} l(y) P_o^m(d\psi) M(dm) \tag{8}$$

which is interpreted as the mean total length of grain boundaries with orientation in L corresponding to germs from S with respect to the reduced Palm distribution P_o^m, M being the mark distribution. Up to a scale factor (8) is the reduced correlation measure K_{pS} defined in [8]. The inner sum is the mark-sum measure.

For an mxn matrix K_{ij} of non-negative elements the following characteristics are defined: the total "energy" D_T

$$D_T = \left[\frac{1}{mn-1}\left(\frac{mn\sum\sum K_{ij}^2}{(\sum\sum K_{ij})^2} - 1\right)\right]^{\frac{1}{2}} \tag{9}$$

and the variability in rows C_A

$$C_A = \left[\frac{1}{n-1}\sum\sum (K'_{ij} - K'_{.j})^2\right]^{\frac{1}{2}} \tag{10}$$

$$K'_{.j} = \frac{1}{n'}\sum_i K'_{ij}, \qquad K'_{ij} = \frac{K_{ij}}{\sum_j K_{ij}}$$

here only non-zero rows of the matrix (their number is $n' \leq n$ assumed that $n' \geq 2$) are included into the evaluation. It is easy to see that $0 \leq D_T$, $C_A \leq 1$ where the upper bound 1 is attained for a matrix with single non-zero value in (9) and for a matrix with a most one non-zero value in each row and column in (10).

Now for $L_j = \langle \beta_j - \Delta, \beta_j + \Delta)$, $j=1,\dots,n$, see (2), and sectors $S_{qq}^{\beta_i\Delta}$, given q

$$K_{ij} = K_{L_j}(S_{qq}^{\beta_i\Delta}), \quad i,j=1,\dots,n \tag{11}$$

is a matrix for which D_T is the degree of total inner anisotropy and C_A is a measure of the correlation between anisotropies of the grain shape and orientation and the spatial distribution of germs.

Let $(M^*, \mathcal{M})$, $(N, \mathcal{N})$ be measurable spaces of realisations of marked point processes with real marks, point processes in R^2, respectively, [10]. Let $\sim$ be an equivalence on M^*: $\psi_1 \sim \psi_2$ if and only if $\varphi(\psi_1) = \varphi(\psi_2)$, where $\varphi(\psi) \in N$ is equal to $\{x_i\}$ for $\psi = \{x_i, m(x_i)\}$. $\tilde{\psi}$ denotes the equivalence class containing ψ. The factor space $\tilde{M}$ of all $\tilde{\psi}$

is endowed with a σ-algebra $\tilde{\mathcal{N}}$ so that $(\tilde{M},\tilde{\mathcal{M}})$ can be identified with $(N,\mathcal{N})$. For a stationary marked point process Ψ let $\Phi=\varphi(\Psi)$ and P_o^m, P_o are Palm distributions of Ψ, Φ, respectively. For $Y\in\tilde{\mathcal{M}}$ let $P_o^M(Y)=\int P_o^m(\bar{Y})M(dm)$ be a measure on $\tilde{\mathcal{M}}$, $\bar{Y}=\{\psi\in M^*,\tilde{\psi}\in Y\}$. Then P_o^M can be identified with P_o.

Consider a marked point process Ψ with the Neymann-Scott point process $\Phi=\varphi(\Psi)$ with intensity λ. A formula $P_o=P*c_o$ [1] holds for the distribution P of Φ, c_o being the Palm distribution of the cluster. According to the above reasoning for bounded $S\in\mathcal{B}^2$ (8) yields:

$$K_L(S)=\int\sum_{\substack{(y,m(y))\\ 0\neq y\in\varphi\cap S}} m(y)P(d\varphi)+\int\sum_{\substack{(y,m(y))\\ 0\neq y\in\varphi\cap S}} m(y)c_o(d\varphi)\;. \qquad (12)$$

Example 2: Let the clusters in the Neymann-Scott process be triplets of points lying in a line with orientation α. The fixed rectangular grains (Fig. 2a) yields the marks in the manner defined above. The matrix (11) is evaluated for n=4, a=5, b=1, r=6, λ=0.001, q=15 (the sector with all information on anisotropy) and $\alpha=\Delta$ fixed (case I) or when α has a uniform random distribution on $\langle 0,\pi)$ (case II). The quantities in Table 2 demonstrate

$$\text{I}\;\begin{pmatrix}10.87 & 0 & 2.17 & 0\\ 0.87 & 0 & 0.17 & 0\\ 0.87 & 0 & 0.17 & 0\\ 0.87 & 0 & 0.17 & 0\end{pmatrix}\qquad \text{II}\;\begin{pmatrix}2.76 & 0.26 & 0.76 & 0.26\\ 0.26 & 2.76 & 0.26 & 0.76\\ 0.76 & 0.26 & 2.76 & 0.26\\ 0.26 & 0.76 & 0.26 & 2.76\end{pmatrix}$$

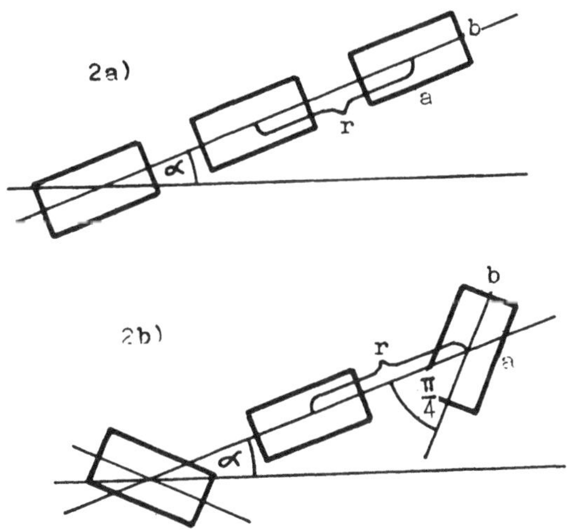

Fig. 2: A typical cluster of the structure in example 2

Table 2: The evaluated parameters D_T, C_A in example 2

	case I	case II
D_T	0.715	0.369
C_A	0.001	0.586

the interpretation of the proposed characteristics D_T, C_A. The meaning of "inner" anisotropy now becomes clear. In the case II it is $D_T > 0$ even if Ψ is an isotropic marked point process.

According to [8] it holds

$$K_L(-S) = K_L^*(S) = \int m \int \sum_{\substack{(y,l(y))\in\Psi \\ 0\neq y\in S}} l(y) P_o^m(d\psi) M(dm) \qquad (13)$$

When taking a cluster from Fig. 2b into example 2 with $\alpha = \Delta$ fixed for $L = \langle \pi/4, \pi/2)$ and $S = S_{qq}^{\beta_1 \Delta}$ then $K_L(S) \neq K_L(-S)$ due to the fact that the rose of directions differs at the opposite ends of the cluster. Therefore in order to detect this property it is necessary that the sectors take orientations round $\langle 0, 2\pi)$. The definitions (9), (10) remain the same.

The estimation of K_{ij} using the realisation of Ψ in a bounded window W requires, for the first step, a special software for an image analyser which measures the point positions and marks. If this is not available the rose of intersections of each grain must be evaluated. For the second step the estimators $k_L(S)$ ($k_L^*(S)$) of $K_L(S)$ ($K_L^*(S)$) are computed as

$$k_L(S) = \sum_{\substack{(x,m),(y,l)\in\Psi\cap W \\ y-x\in S}}^{\neq} \frac{1}{\nu_2(W_x \cap W_y)} \qquad (14)$$

$$k_L^*(S) = \sum_{\substack{(x,m),(y,l)\in\Psi\cap W \\ y-x\in S}}^{\neq} \frac{m}{\nu_2(W_x \cap W_y)}$$

where ν_2 is the Lebesque measure on $\mathbb{R}^2$. The unbiasedness of these estimators follows from standard arguments using the theory of mark-sum measures [8]. In any case, only sectors around $\langle 0, \pi)$ are used and the computation of both k_L and k_L^* yields the complete information as a result of (13).

References

[1] Ambartzumian, R.V.: On an equation for stationary point processes. (Russ.) Dokl.Akad.Nauk Armjanskoj SSR 42 (1966), 141-147.

[2] Hanisch, K.H.: Reduction of n-th moment measures and the special case of the third moment measure of stationary and isotropic planar point processes. Math.Operationforsch.Statist. 14 (1983), 421-436.

[3] Hanisch, K.H., Stoyan, D.: Once more on orientations in point processes. Elektron.Inf.und Kyber. 20 (1984), 279-284.

[4] Hilliard, J.E.: Specification and measurement of microstructural anisotropy. Trans.Metall.Soc.AIME 224 (1962), 1201-1211.

[5] Kshirsagar, A.M.: Multivariate Analysis. Marcel Dekker, Inc. New York 1972.

[6] Ohser, J., Stoyan, D.: On the second order and orientation anylysis of planar stationary point process. Biometrical J. 23 (1981), 523-533.

[7] Ripley, B.D.: The second-order analysis of stationary point processes. J.Appl.Prob. 13 (1976), 255-266.

[8] Stoyan, D.: On correlations of marked point processes. Math.Nachrichten 116 (1984), 197-207.

[9] Stoyan, D., Mecke, J., Pohlmann, S.: Formulas for stationary planar fibre processes II: Partially oriented fibre systems. Math.Oper.Statist. 11 (1980), 281-286.

[10] Stoyan, D., Kendall, W.S., Mecke, J.: Stochastic geometry and its applications, Akademie Verlag, Berlin 1987.

[11] Wiencek, K., Hougardy, H.: Description of the homogeneity of particle arrangement. Acta Stereol. 6/1 (1987), 69-74.

GEOMETRIC COVARIOGRAMS FOR PAIRS OF PARALLEL j-PLANES

Werner Nagel[1]

1. Motivation

The geometric covariogram of a set as it was introduced and studied by Matheron (see [3]) is an established "second order quantity" in integral and stochastic geometry. It is closely related to the co-occurence matrices for digital binary images which are an important tool in texture analysis. The notions of a covariogram and of a co-occurence matrix have been developed independently in the two fields.

There are several open problems concerning the amount of information about a set which is contained in its covariogram. For example, is it possible to determine the Minkowski-functionals of a compact convex set or even the set itself from the covariogram? From Matheron's work we know that the volume and the surface content are given by the value of the covariogram in the origin and by the first derivatives around the origin respectively. Further, Matheron derived a formula for the distribution function of the chord lengths of lines sectioning a convex set parallel to a fixed direction. These distributions provide an information about the set which is weaker than the Radon transform (of the indicator function of the set) but stronger than the usual chord length distribution over all directions of lines.

Current research in stereology concerning parallel sections such as the disector and its modifications (see [1]) motivates the investigation of generalizations of the geometric covariogram, that is functions which depend on observations on pairs of parallel lines or planes instead of pairs of points.

In the present paper the Radon transform is applied to the geometric (point) covariogram to relate it to some of these generalized covariograms. This result can be interpreted as a second order analogue of a Crofton formula in classical integral geometry.

2. The geometric covariogram and its generalization

Let A be a Borel set of the n-dimensional Euklidean space R^n. Its indicator function is denoted by 1_A.

Definition 1 (Matheron):
The function $K_A: R^n \longrightarrow [0,\infty]$ given by

$$K_A(h) = \int 1_A(x) \cdot 1_A(x+h)\, dx, \quad h \in R^n, \tag{1}$$

is called the <u>geometric covariogram</u> of A.

Clearly,

$$K_A(h) = V(A \cap (A-h)) \tag{2}$$

where V denotes n-dimensional volume. Instead of the indicator function one can use the Euler-Poincaré-characteristic $C_0^0(A \cap \{x\}) = 1_A(x)$. Thus

[1]Friedrich-Schiller University, Dept. of Mathematics, Jena, UHH, 17.OG, DDR-6900

$$K_A(h) = \int C_0^0(A \cap \{x\}).C_0^0\,(A \cap \{x+h\})dx. \tag{3}$$

Let L denote a fixed j-dimensional linear subspace of R^n and $L^\perp$ its orthogonal complement for j = 0, ..., n. The orthogonal projection onto $L^\perp$ is denoted by $\pi(.,L^\perp)$. The Minkowski-functionals (quermassintegrals, curvature integrals) in the j-dimensional space are denoted by C_k^j, k = 0, ..., j. In particular, C_j^j is the j-dimensional volume and C_0^j is the Euler-Poincaré-characteristic. (Note that this notation differs from that one used for W_i^k in [3]. They are related by

$$C_k^j = \binom{j}{k}\, b_{j-k}^{-1}\, W_{j-k}^j\ , \quad k = 0, \ldots, j.$$

where b_i denotes the volume of the i-dimensional unit sphere.) Consider only such sets A for which all the functionals C_k^j are defined (such as convex sets, finite unions of convex sets, sets of positive reach, cf. [5].)

Definition 2:

The function $K_{A,L}^j: L \longrightarrow R^1$ given by

$$K_{A,L}^j(h) = \int_{L^\perp} C_n^j(A \cap (L+x)).C_0^j(A \cap (L+x+h))dx, \qquad h \in L^\perp, \tag{4}$$

is called the j-plane covariogram of A in direction L.

For the geometric covariogram of definition 1 we can write $K_A = K_{A,0}^0$. If A is a convex set then the integrand in (4) is either 1 (if and only if both the planes L+x and L+x+h hit A) or 0. Hence we obtain

Proposition:

If A is a convex set then

$$K_{A,L}^j = K^j_{\pi(A,L^\perp),L} = K_{\pi(A,L^\perp)} \tag{5}$$

where these functions are all defined on the (n-j)-dimensional subspace $L^\perp$.

Thus the j-plane covariogram of a convex set equals the geometric covariogram of the projected set. Evidently, the volume of the projection is $C_{n-j}^{n-j}(\pi(A,L^\perp)) = K_{A,L}^j(0)$. Its surface content 2 $C_{n-j-1}^{n-j}(\pi(A,L^\perp))$ one gets by differentiation of $K_{A,L}^j$, using Matheron's formula ([3], Theorem 4.3.1). An integration over all j-dimensional subspaces L ("all directions") and the application of Kubota's recursion formula of integral geometry yields $C_{n-j}^n(A)$ (see [3], (4.1.3), (4.1.5)) and $C_{n-j-1}^n(A)$ (see (4.1.6)). Without giving the details here, it should be clear that for convex sets A all the Minkowski-functionals $C_k^n(A)$ can be determined by (n-k)-plane covariograms or by (n-k-1)-plane covariograms resp., k = 0, ..., n. A survey of the results for n = 2 and n = 3 is given in §4.

Definition 2 immediately suggests that we consider not only the product of the Euler-Poincaré-characteristics but also products of the other Minkowski-functionals. This leads to

Definition 3:

The function $K_{A,L}^{j,k}: L \longrightarrow R^1$ given by

$$K^{j,k}_{A,L}(h) = \int_{L^\perp} C^j_k(A \cap (L+x)).C^j_k(A \cap (L+x+h))dx, \quad h \in L^\perp, \tag{6}$$

is called the <u>k-th j-plane covariogram of A in direction L.</u>

Clearly, $K^j_{A,L} = K^{j,o}_{A,L}$.

3. Stereological formulae for covariograms

It is a stereological problem to investigate relationships between the covariograms of definition 3 for several dimensions j of the (pairs of parallel) section planes. A first result can be derived by applying the j-dimensional Radon transform to the geometric covariogram.

Theorem:

Let A be a Borel set in R^n, L a fixed j-dimensional subspace and $L^\perp$ its orthogonal complement, j = 1, ..., n-1. Then for all $z \in L^\perp$ we have

$$K^{j,j}_{A,L}(z) = \int_{L+z} K_A(h)\; dh \tag{7}$$

that is the integral over the j-plane L+z of the geometric covariogram equals the value of the j-th (volume) j-plane covariogram at z for the same subspace L.

<u>Proof</u>: From the definition of K_A, Fubini's theorem and a translative integral geometric formula (similar to the principal kinematic formula) for volumes in the j-dimensional space we derive

$$\begin{aligned}\int_{L+z} K_A(h)dh &= \int_{L+z} C^n_n(A \cap (A-h))\; dh \\ &= \int_L C^n_n(A \cap (A-z-t))\; dt \\ &= \int_L \int_{L^\perp} C^j_j(A \cap (A-z-t) \cap (L+y))\; dy\; dt \\ &= \int_{L^\perp} C^j_j(A \cap (L+y)).C^j_j((A-z) \cap (L+y))\; dy \\ &= K^j_{A,L}(z).\end{aligned}$$

This proof works only for point covariances and it can <u>not</u> be used to derive formulae which link $K^{j,j}_{A,L}$ and $K^{i,i}_{A,L}$ for $i \neq j$, and it is also not applicable for $K^{j,k}_{A,L}$, $k < j$.

The theory of the Radon transform allows us to give an inverse formula to (7). It can be obtained from theorem 2.27 in I.1.6 of [2]. Thus one has the possibility to determine for any L the j-th (i.e. volume-) j-plane covariogram from the geometric (point-) covariogram and vice versa. This result should be taken into consideration in image processing and statistics.

Incidentally, formula (7) can be used to give an integral expression for the second moment of the chord length distribution and also to derive a classical formula which relates this second moment to the (1-n)-th moment of the distance of two random points (see [4], p.238). To see this, let A be a compact <u>convex</u> set, put j = 1, z = 0, and integrate both sides of (7) over all lines through the origin,

$$\int_{L}\int_{L} K_A(h)\ dh\ dL = \int_{L} K^{1,1}_{A,L}(0)\ dL = \int_{L}\int_{L^\perp} C^1_1(A\cap(L+z))^2\ dz\ dL \tag{8}$$

Here L is the set of all 1-dimensional subspaces of R^n and dL the element of the rotation invariant measure on L with $dL = O_n/2$, and O_n is the surface of the n-dimensional unit sphere. Up to a constant, the right hand side is the second moment of the chord length distribution for A, and this can easily be expressed by the geometric covariogram as (8) shows. Substituting the definition of $K_A(h)$ and applying the Blaschke-Petkantschin formula (cf. [4], (12.23)) to the left side one obtains

$$\int_{A}\int_{A} \|x_1-x_2\|^{1-n}\ dx_1\ dx_2 = \int_{L}\int_{L^\perp} (C^1_1(A\cap(L+z)))^2\ dz\ dL \tag{9}$$

Note that C^1_1 is the length and dz dL is the element of a motion-invariant measure on the set of all lines in R^n.

4. Summary of the results for R^2 and R^3

For the two-dimensional space R^2 we have defined three covariograms: For pairs of points

a) $K_A(h) = K^0_{A,0}(h)$, the area $F(A\cap(A-h))$ for $h \in R^2$.

For pairs of lines parallel to a fixed line L,

b) $K^1_{A,L}(h) = K^{1,0}_{A,L}(h)$ ... the translative integral (w.r.t. $x \in L^\perp$) of the product of the Euler-Poincaré-characteristics $C^1_0(A\cap(L+x)) \cdot C^1_0(A\cap(L+x+h))$, $h \in L^\perp$. If A is a convex set then $K^1_{A,L}(h) = l(\pi(A,L^\perp)) - \|h\|$, which is the length of the intersections of the projections of A and of A-h. This length can be determined by a first derivative of K_A.

c) $K^{1,1}_{A,L}(h)$, the translative integral (w.r.t. $x \in L^\perp$) of the product of the chord lengths $l(A\cap(L+x)).l(A\cap(L+x+h))$, $h \in L^\perp$. From the theorem it follows that $K^{1,1}_{A,L}$ can be derived from K_A and vice versa (see figure 1).

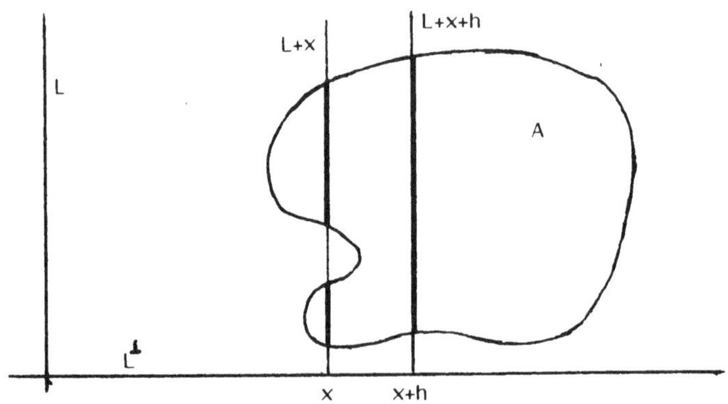

Figure 1: Illustration of the integrand in definition 3 for j=1 (section lines) in the two-dimensional space.

These results show that for convex sets A in R^2 the line covariograms do not provide more information than the geometric (point-) covariogram K_A.

In the three-dimensional space R^3 we have, for pairs of points

a) $K_A(h) = K^o_{A,0}(h)$, the volume $V(A \cap (A-h))$ for $h \in R^3$

For pairs of lines parallel to a fixed line L,

b) $K^1_{A,L}(h) = K^{1,0}_{A,L}(h)$ is literally the same as in b) for R^2. If A is convex then, according to (5), $K^1_{A,L}(h) = F(\pi(A,L^\perp) \cap \pi(A-h,L^\perp))$, which is the area of the intersections of the orthogonal projections of A and of A-h onto $L^\perp$ (see figure 2).

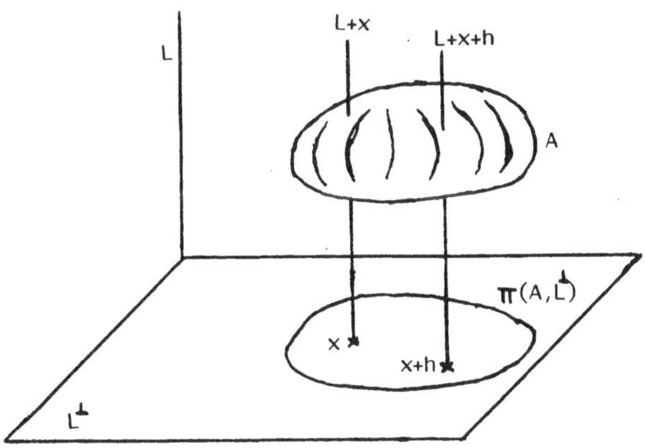

Figure 2: Illustration of (5) for j=1 (section lines) in the three-dimensional space.

c) $K^{1,1}_{A,L}(h)$, the translative integral (w.r.t. $x \in L^\perp$) of the product of the chord lengths $l(A \cap (L+x).l(A \cap (L+x+h))$, $h \in L^\perp$. By the theorem $K^{1,1}_{A,L}$ can be given by a 1-dimensional Radon-transform of K_A.

For pairs of planes parallel to a fixed plane L,

d) $K^2_{A,L}(h) = K^{2,0}_{A,L}(h)$... the translative integral (w.r.t. $x \in L^\perp$) of the product of the Euler-Poincaré-characteristics $C^2_0(A \cap (L+x)) \cdot C^2_0(A \cap (L+x+h))$, $h \in L^\perp$. If A is convex then $K^2_{A,L}(h) = l(\pi(A,L^\perp)) - \|h\|$, which is the length of the intersection of the projections of A and of (A-h) onto $L^\perp$. Since $l(\pi(A,L^\perp))$ equals the breadth of A in direction $L^\perp$ and the support of the function K_A is supp $K_A = A \oplus (-A)$, the values of $K^2_{A,L}$ can be determined from K_A. Further, for convex A the function $K^2_{A,L}$ can also be determined by derivatives of $K^1_{A,L}$.

e) $K^{2,1}_{A,L}(h)$, one quarter of the translative integral (w.r.t. $x \in L^\perp$) of the product of boundary lengths $l(\partial(A \cap (L+x))).l(\partial(A \cap (L+x+h)))$, $h \in L^\perp$.

f) $K^{2,2}_{A,L}(h)$, the translative integral (w.r.t. $x \in L^\perp$) of the product of areas $F(A \cap (L+x)).F(A \cap (L+x+h))$, $h \in L^\perp$. This can be obtained from K_A by a Radon transform as described in the theorem.

Thus, if A is convex only for $K^{1,0}_{A,L}$ and for the first 2-plane covariogram $K^{2,1}_{A,L}$ do we have no formulae to determine the function from K_A. Note that formulae which relate directly the line covariograms (in b) and c)) to the 2-plane covariograms are also unavailable.

5. Literature

[1] Gundersen, H.J.G.: Stereology of arbitrary particles. A review of unbiased number and size estimators and the presentation of some new ones. J. Microscopy **143** (1986), 3-45.

[2] Helgason, S.: Groups and Geometric Analysis. Academic Press, 1984. (Russ. transl.: Mir, Moskva 1987).

[3] Matheron, G.: Random Sets and Integral Geometry. Wiley, New York 1975.

[4] Santaló, L.A.: Integral Geometry and Geometric Probability. Addison-Wesley, Reading 1976.

[5] Zähle, M.: Curvature measures and random sets I. Math. Nachr. **119** (1984), 327-339.

[6] Stoyan, D., W.S.Kendall and J. Mecke: Stochastic geometry and its applications. J. Wiley/Akademie-Verlag, Chichester/Berlin 1986

X-RAY PICTURES OF CONVEX SETS

Horst Martini [1)]

We shall give a survey on results from geometric convexity which might be informative for stereologists.
As is well-known, the numerical reconstruction of the density distribution of an object $Q \subset R^d$ from X-ray pictures is the subject of computerized tomography (cf. Natterer [20]). The density function f of Q cannot be reconstructed if, for instance, only a finite number of X-ray pictures of Q is taken into consideration. But the restriction to convex sets of homogeneous density allows some progress and gives interesting uniqueness and reconstruction problems. It should be noted that also this restricted viewpoint has applications (see e.g. Das and Boerner [2]).
Clearly, the mentioned results are only selected from those contained in the cited papers.

§ 1 NOTATION

Let R^d ($d \geq 2$) denote the d-dimensional Euclidean vector space with inner product $\langle \cdot , \cdot \rangle$ and unit sphere $S^{d-1} := \{u \in R^d \mid \langle u,u \rangle = 1\}$. We write K^d for the set of convex bodies, i.e. compact, convex subsets of R^d with interior points.
Since we generally assume homogeneous density of $K \in K^d$, it suffices to use a purely geometric view, in which the X-ray picture of K from a direction $u \in S^{d-1}$ is presented by a function on $H_u := \{z \in R^d \mid \langle z,u \rangle = 0\}$, whose value at z is the length of the intersection of K and the line through z with direction u. On the other hand, the X-ray picture of K from a point $p \in R^d$ is given by a function on the sphere $p + S^{d-1}$, whose value at $p + u$ is the length of the intersection of K and the line through p parallel to $u \in S^{d-1}$

§ 2 UNIQUENESS RESULTS

The following problem of Hammer [13] outlines our considerations: Suppose we are given the lengths of the intersections of $K \in K^2$

1) Sektion Mathematik der Pädagogischen Hochschule
Wigardstraße 17
DDR - 8060 Dresden

and all lines in directions $u_1,\dots,u_n$ or issuing from points $p_1,\dots,p_m$ not from K. How many directions or points must be used to permit the exact reconstruction of K ?

In 1962, Giering [9] proved that three directions, depending on K, may be choosen so that by the X-ray pictures in those directions K is uniquely determined. Further it was shown that in general two directions are not sufficient. (For a new proof of Giering's theorem see e.g. Gardner [6].)

In [8] Gardner and McMullen proved that independently from $K \in K^d$ X-ray pictures in four directions, whose slopes with respect to some coordinate system have a transcendental cross-ratio, distinguish between all convex bodies from K^d, and that for each direction triple there exist two distinct convex bodies having the same X-ray pictures in these directions. (Note that convex bodies with the same X-ray pictures as a given one are called "ghosts".)

Volčič [25] and Longinetti [16] investigated, how heavy non-uniqueness in R^2 can be, when uniqueness of the reconstruction is not assured. In particular, Longinetti gave an upper bound on the L_2-distance between any two planar solutions to the Hammer problem with respect to $n \geq 3$ directions.

Besides results of Giering [10], who constructed pairs of triangles and quadrangles with equal X-ray pictures in two directions, seemingly no deeper results on convex ghosts are known. This led Volčič and Zamfirescu [30] to show that, in the sense of Baire category results, most convex bodies in R^d $(d \geq 2)$ are not ghosts with respect to two directions.

Convex polytopes, i.e. convex bodies in R^d with a finite number of extreme points, are uniquely determined by a nearly arbitrary pair of X-ray pictures. At least in 3-space, the same holds for non-convex polyhedra (Golubyatnikov [11],[12]). Analogously, a nearly arbitrary pair of directions uniquely determines a smooth convex body of revolution in R^3, as shown by Rykov [22]. (Here and above "nearly arbitrary" means that only unimportant exceptions exist.)

Some of the mentioned results have analogies for point sources. Besides other results Falconer [5] showed that, with a few exceptions, a given convex set $K \subset R^2$ can be distinguished from any other one by the X-ray pictures taken from $p_1, p_2 \in R^2$ such that the line through p_1 and p_2 meets the interior of K. This result is even reconstructive (cf. § 3), whereas its uniqueness part was reproved by Gardner [6]. Further Falconer confirmed that there is only a finite number of ghosts for point sources.

With respect to the projective plane, i.e. the completion of R^2 by the line at infinity, Volčič [26] observed that X-ray pictures from four points, no three of which are collinear, or from three non-collinear points not contained in the X-rayed sets, distinguish between all plane convex bodies. He also gave a projective version of Giering's basic theorem, namely: There exist three suitable points on a line of the projective plane such that the corresponding X-ray pictures distinguish the plane convex body K from any other one (cf. [28]). Volčič further obtained interesting uniqueness theorems for mixed X-ray pictures of convex bodies in the Euclidean plane, i.e. by using directions and point sources for the same object (see [27]). For generalizations in this direction, partially also with reconstruction methods, the reader is referred to the papers [4], [7], and [23] of Falconer, Gardner, and Zuccheri.

§ 3 RECONSTRUCTION RESULTS

In [24] and [1] it was shown that the reconstruction problems corresponding to the uniqueness results of Gardner-McMullen (for directions) and Volčič (for point sources) are well-posed.
Chang and Chow [3] as well as Volčič [29] gave special methods for reconstructing certain subclasses of planar bodies from X-ray pictures in two directions.
In a deep paper of Kölzow, Kuba and Volčič [14] an algorithm is derived which allows the reconstruction of some $K \in K^2$ to any required degree of accuracy from X-ray pictures in finitely many directions. Error estimates are given, and after small modification the algorithm can be applied to reconstruct K even from discrete projectional data.
Algorithmical investigations of Michelacci [19] continue the investigations of Volčič in [28].

§ 4 CONCLUDING REMARKS

Additionally, results of Reed et al. [21] and Kuba-Volčič [15] should be mentioned, where those plane sets of finite 2-dimensional Lebesgue measure are characterized which can be uniquely determined by X-ray pictures from two directions.
We finish by attracting reader's attention to another type of investigations. Nevertheless, these investigations refer to stereological aims by using X-ray pictures, too.
It is clear, that the volume V(K) of $K \in K^d$ ($d \geq 2$) can be obtained by an integration process with respect to one arbitrary X-ray pic-

ture of that convex body. But, denoting by $\overline{V}_{d-1}(K,u)$ the area of the orthogonal projection of K in H_u and by $\underline{V}_1(K,u)$ the maximal chord length of K in direction u, a d-simplex S satisfies

$$\overline{V}_{d-1}(K,u)\cdot\underline{V}_1(S,u) = d\ V(S) \qquad (1)$$

for each direction $u \in S^{d-1}$ (Martini [17]). Thus one has a convenient approach to V(S) from one arbitrary X-ray picture without integration methods, only using the measures $\overline{V}_{d-1}(S,u)$ and $\underline{V}_1(S,u)$. It is also obvious that a d-ellipsoid E satisfies

$$\overline{V}_{d-1}(E,u)\cdot\underline{V}_1(E,u) = \frac{2\omega_{d-1}}{\omega_d} V(E) \qquad (2)$$

for each direction u, where ω_d is written for the volume of the unit d-ball.

The question for further classes of convex bodies in R^d with a constant product $\overline{V}_{d-1}(\cdot,u)\cdot\underline{V}_1(\cdot,u)$ over all directions could be answered completely only for polytopes in K^d: A convex d-polytope ($d \geq 3$) has this property if and only if it is a simplex (cf. [17]). Moreover, $K \in K^d$ is already a simplex if (1) holds for all directions from an arbitrary triple of great (d-2)-subspheres on S^{d-1}, as shown in [18].

(In R^2 the product $\overline{V}_{d-1}(\cdot,u)\cdot\underline{V}_1(\cdot,u)$ is constant over S^1 precisely for those convex domains K whose difference body $DK = \{x \in R^2 \mid x = x_1 - x_2 \text{ with } x_1, x_2 \in K\}$ is bounded by a so-called Radon curve.)

REFERENCES

[1] Bianchi, G., A. Volčič: Hammer's X-ray problem is well-posed Ann. di Mat. Pura e Appl. (to appear).

[2] Das, Y., W. M. Boerner: On Radon target shape estimation using algorithms for reconstruction from projections. IEEE Transactions AP 26 (1978), 274-279.

[3] Chang, S. K., C. K. Chow: The reconstruction of three-dimensional objects from two orthogonal projections and its applications to cardiac cineangiography. IEEE Trans. Comput. 22 (1973), 18-28.

[4] Falconer, K. J.: Hammer's X-ray problem and the stable manifold theorem. J. London Math. Soc. (2) 28 (1983), 149-160.

[5] Falconer, K. J.: X-ray problems for point sources. Proc. London Math. Soc. (3) 46 (1983), 241-262.

[6] Gardner, R. J.: Symmetrals and X-rays of planar convex bodies. Arch. Math. 41 (1983), 183-189.

[7] Gardner, R. J.: Chord functions of convex bodies. J. London Math. Soc. (2) 36 (1987), 314-326.

[8] Gardner, R. J., P. McMullen: On Hammer's X-ray problem. J. London Math. Soc. (2) 21 (1980), 171-175.

[9] Giering, O.: Bestimmung von Eibereichen und Eikörpern durch Steiner-Symmetrisierungen. Sber. Bayer. Akad. Wiss., München, Math.-Nat. Kl. (1962), 225-253.

[10] Giering, O.: Drei- und Viereckspaare, für deren Drei- und Vierecke jeweils zwei Steiner-Symmetrisierungen übereinstimmen. Elem. Math. 40 (1985), 1-10.

[11] Golubyatnikov, V. P.: On the tomography of polyhedra (Russian). In: Questions on correctedness in inverse problems of mathematical physics. Computer Center Novosibirsk 1982, 75-76.

[12] Golubyatnikov, V. P.: Questions on unique solutions of inverse problems with respect to the reconstruction of shapes of bodies from their projections (Russian). Preprint no. 3, Sibirsk. Otdel. Akad. Nauk SSSR, Math. Inst., Novosibirsk 1988 (26 pp.).

[13] Hammer, P. C.: Problem 2. In: Convexity, Proc. Sympos. Pure Math. 7 (Amer. Math. Soc., Providence, 1963), 498-499.

[14] Kölzow, D., A. Kuba, A. Volčič: An algorithm for reconstructing convex bodies from their projections. Discrete Comput. Geom. (to appear).

[15] Kuba, A., A. Volčič: Characterization of measurable plane sets which are reconstructable from their two projections. Inverse Problems 4 (1988), 513-527.

[16] Longinetti, M.: An isoperimetric inequality for convex polygons and convex sets with the same symmetrals. Geom. Dedicata 20 (1986), 27-41.

[17] Martini, H.: Convex polytopes whose projection bodies and difference sets are polars. Discrete Comput. Geom. (to appear).

[18] Martini, H.: Determining classes of convex bodies by restricted sets of Steiner symmetrizations. Geom. Dedicata (to appear).

[19] Michelacci, G.: An iterative algorithm for reconstructing inscribed triangles. Manuscript.

[20] Natterer, F.: The Mathematics of Computerized Tomography. Teubner Stuttgart, Wiley New York et al., 1986.

[21] Reed, J. A., L. A. Shepp, P. C. Fishburn, J. C. Lagarias: Sets uniquely determined by projections. Technical Memorandum, AT & T Bell Laboratories, 1986.

[22] Rykov, T. L.: Reconstruction of a body of revolution from two X-ray pictures (Russian). In: Questions on correctedness in inverse problems of mathematical physics. Computer Center Novosibirsk 1982, 101-103.

[23] Zuccheri, L.: Approximation of convex bodies from chord functions. Quad. Mat. Univ. Trieste 132, 1987 (21 pp.).

[24] Volčič, A.: Well-posedness of the Gardner-McMullen reconstruction problem. Proc. Conf. Measure Theory, Oberwolfach 1983. Ed. by D. Kölzow and D. Maharm-Stone, Lecture Notes Math. 1089, Springer, Berlin et al., 1984.

[25] Volčič, A.: Ghost convex bodies. Boll. U. M. I. (6) 4 - A (1985), 287-292.

[26] Volčič, A.: A three-point solution to Hammer's X-ray problem. J. London Math. Soc. (2) 34 (1986), 349-359.

[27] Volčič, A.: Uniqueness theorems for mixed X-ray problems of Hammer type. Preprint (DFG Research Project Reconstruction Ko 506/8-1, directed by D. Kölzow, Univ. of Erlangen).

[28] Volčič, A.: Tomography of convex bodies and inscribed poly-

gons. Ric. di Mat. (to appear).

[29] Volčič, A.: On a class of plane sets which can be reconstructed from two Steiner symmetrizations. Manuscript.

[30] Volčič, A., T. Zamfirescu: Ghosts are scarce. J. London Math. Soc. (to appear).

ON BASIC NOTIONS OF FUZZY SET THEORY AND SOME IDEAS FOR THEIR APPLICATION IN IMAGE PROCESSING

H. Bandemer, A. Kraut, W. Näther)[1]

1. Basic notions

A subset of a basing set U (the universe of discourse) can be characterized by its characteristic function

$$\forall\, x \in U : m_A(x) = 1 \text{ if } x \in A, \text{ and } 0 \text{ otherwise.} \tag{1.1}$$

As is well known, every mathematical statement can be expressed by the set, where it holds. Hence mathematics, in principle, can be traced back to manipulations with sets.

To meet the requirements of automatic control in man-machine-systems, where information and decisions are formulated by verbal statements, Zadeh ([14], [15]) suggested, some twenty years ago, to allow the elements of U to assume gradual memberships (e.g. between zero and one) to different subsets. In analogy to (1.1) he defined a membership function (m.f.) by

$$m_A : U \text{ --> } [0,1] \tag{1.2}$$

and called the so specified subset A a fuzzy set (f.s.).

If we identify the membership value 1 with the level "black" and the value 0 with "white", every f.s. can be represented by a grey-tone picture over U. Manipulating with f.s.'s then corresponds to image processing. Hence the rules for handling f.s.'s are of interest also in this branch of science.

Whereas for common, in this context named crisp, sets intersection and union are defined uniquely, for fuzzy sets there are several possibilities for the definition of this connections (cf. e.g. Dubois/Prade [6]; Bandemer/Gottwald [2]). The most frequently used version is

$$C = A \cap B \;:\; m_C = \min\{ m_A(x),\, m_B(x) \} \tag{1.3}$$

$$D = A \cup B \;:\; m_D = \max\{ m_A(x),\, m_B(x) \}. \tag{1.4}$$

The complement $\tilde{A}$ is usually defined by

$$m_{\tilde{A}}(x) = 1 - m_A(x). \tag{1.5}$$

Note that the statements for a crisp set: $A \cap \tilde{A} = \emptyset$ and $A \cup \tilde{A} = U$ do not hold for any fuzzy non-crisp set.

To transfer the fuzziness from the domain into the range of any function, transformation or mapping, say f, Zadeh suggested his extension principle

$$m_{f(A)}(y) = \sup_{x\,:\,f(x)=y} m_A(x). \tag{1.6}$$

)[1] Bergakademie Freiberg, Sektion Mathematik, PF 47

If several arguments operate independently their corresponding fuzzy sets are, at first, to be intersected. E.g. if $f(x,y) = x + y$ then $f(A,B) = A \oplus B$ with

$$m_{A\oplus B}(z) = \sup_{x,y\,:\,x+y=z} \min \{ m_A(x), m_B(y) \}. \tag{1.7}$$

This defines an addition of f.s.'s. Obviously, multiplication and division can be defined likewise.
For a numerical handling of f.s.'s it is essential to consider α-cuts, i.e. crisp sets A_α, defined by

$$A_\alpha = \{ x : m_A(x) \geq \alpha \} \tag{1.8}$$

(also strong α-cuts $A_{>\alpha}$ with the strict sign $>$ in (1.8)).
From the set of α-cuts the f.s. A can be reconstructed using

$$m_A(x) = \sup_{\alpha \in (0,1]} \alpha \cdot m_{A_\alpha}(x) \tag{1.9}$$

where m_{A_α} is the characteristic function of A_α. Obviously, the α-cuts are monotonic,

$$\forall \alpha_1, \alpha_2 \in (0,1] : \quad \alpha_1 < \alpha_2 \; \text{-->} \; A_{\alpha_1} \supseteq A_{\alpha_2}. \tag{1.10}$$

Especially $A_{>0}$ =: supp A is called the support of A.
The convexity of a f.s. A is defined by

$$\forall x,y \in U \quad \forall \lambda \in [0,1] : m_A(\lambda x + (1-\lambda)y) \geq \min \{ m_A(x), m_A(y) \} \tag{1.11}$$

which corresponds to the property that each α-cut is a convex crisp set.
In analogy to the definition of cardinality for crisp sets we can introduce

$$\text{card } A = \sum_{x \in U} m_A(x) \quad \text{if U is finite} \tag{1.12}$$

and

$$\text{card } A = \int_U m_A(x)\, dx \quad \text{if U is a continuum} \tag{1.13}$$

where U and m_A are measureable with respect to the used integral notion.
Now we are ready to consider mathematical structures in Euclidean spaces, i.e. $U = \mathbf{R}^d$.
We will call a f.s. G with $m_G : \mathbf{R}^d \text{-->} [0,1]$ a well-shaped (d-dimensional) fuzzy region, if all α-cuts are (finite) Lebesgue measurable, i.e.

$$\forall \alpha \in (0,1] \quad \exists \; \text{card}_d \, G_\alpha < \infty \tag{1.14}$$

where card_d means the d-dimensional volume. If $d > 1$ then we demand further

$$\forall \alpha \in (0,1] \quad \exists \; \text{card}_{d-1} \partial G_\alpha < \infty \tag{1.15}$$

where ∂G_α means the surface of G and card_{d-1} is the (d-1)-dimensional volume of this surface. Other definitions of a fuzzy region are possible and used (see Section 3).
Usually we introduce such regions by specifying a core, i.e. G_1, and a

fuzzy surrounding, where $m_G(x) < 1$. In this manner a (one-dimensional) <u>fuzzy number</u> is specified by a crisp number as core, say x_0, and two functions m^L, defined for $x < x_0$, and m^R, for $x > x_0$, monotonically decreasing for increasing distance from x_0. This LR-representation is favourable for an easy numerical computation, because it reduces arithmetic operations with fuzzy numbers to operations with parameters of m^L and m^R.

Two applications in image processing are sketched in the following: A more principal one in Section 2 and a more detailed one in Section 3. But previously it seems to be useful to sketch the main differences between fuzzy-set-theoretic and probabilistic approach. A probabilistic approach describes (by a certain distribution) the set of potential outcomes. An observation result is vague only before experimentation but is a determined object after sampling. In a fuzzy-set-theoretic approach the concrete observed 'picture' is furnished with a certain vagueness (e.g. grey-tone hull) which stems either from the picture itself (if it has vague boundaries, e.g. as a X-ray photo) or from our description of the observed object (e.g. natural colour or intensities by so-called linguistic expressions).

Note that, on a higher level of modelling, also fuzzy objects can occur following a random mechanism (e.g. random grey-tone pictures, see Puri / Ralescu [11]). Note furher, that the probabilistic approach can be weakened if non-additive measures are admitted for evaluation of potential (fuzzy or non-fuzzy) outcomes. This leads to another field of fuzzy theory, to so-called fuzzy measures as belief, plausibility or possibility (see, e.g. Dubois / Prade [6]).

2. Grey-tone image processing with grey-tone structural elements

Image processing for binary pictures with binary structural elements is a well-developed field, from the point of theory as well as from the point of application (see Serra [9]). Especially for investigations of statistical characteristics of random sets (e.g. covariance structure, granulometry) suitable transformations of the pictures are necessary, which go back to a few basic transformations, such as <u>dilatation</u> and <u>erosion</u> of a (crisp) set A by a (crisp) structural element B, i.e.

$$\text{dil } A = A \oplus \hat{B} \tag{2.1}$$

$$\text{eros } A = A \ominus B =: \widetilde{(\tilde{A} \oplus \hat{B})} \tag{2.2}$$

where $\tilde{X}$ is the complement of the set X, $\hat{X} = \{-x \mid x \in X\}$ is the reflection of X and $\oplus$ stands for Minkowski-addition, i.e.

$$X \oplus Y = \{ z=x+y \mid x \in X, y \in Y \} . \tag{2.3}$$

Goetcherian [8] has investigated grey-tone pictures with binary structural elements. In our note also grey-tone structural elements are admitted. We will interpret both, the grey-tone picture and the grey-tone structural element as fuzzy sets as introduced in section 1.

Note that the addition between fuzzy sets (see (1.7)) is a direct generalization of Minkowski-addition (insert crisp sets into (1.7), and you find formula (2.3)). Thus, we define dilatation and erosion with the extended operation $\oplus$ in the same manner as in (2.1) and (2.2), where the reflection $\hat{X}$ of a f.s. X is given by $m_{\hat{X}}(x) = m_X(-x)$.

Let us mention that in Serra's morphology of functions (see [12]) the m.f. m_A of a f.s. A is associated with the so-called umbra $Um(a) = \{(x,y) \mid y \leq m_A(x)\}$. Then dil and eros of a (say two-dimensional) function is introduced as the usual dil- and eros- operation for its (three-dimensional) but crisp umbra. Our algorithm works in the original space and requires no increase of dimension .

To get some feeling how dil A and eros A work, we consider, at first, one-dimensional f.s.'s. For illustration see Fig.1.

Whereas the examples 1 to 7 meet the popular meaning of dilatation and erosion, the examples 8 and 9 lead to new effects which have no analogue in binary image processing: Dilatation of a picture with a proper grey-pixel at level α cuts the tops of the picture which are higher than α. Erosion with such an α-grey-pixel keeps the tops at level higher than $1-\alpha$ and fills the valleys up to this level. The grey A - grey B examples 6 and 7 use a fuzzy zero as structural element. Note that dil and eros with a crisp zero does not change the picture. Operating with a fuzzy zero B shows, however, some change. But dil does not increase the pure black area. Only the proper grey (and white) values can increase. Eros does not increase the pure white area, the grey (and black) values are damped but not erased. Thus, dil and eros with fuzzy zero only change, in some sense, the contrast of the picture.

In a next step we will formulate dil A and eros A as algorithms for image processing, i.e. we consider discretized pictures and discretized structural elements. The algorithms follow directly from the basic (addition) formula (1.7). We get the grey-tone value of dil A [eros A] for the picture element z_i in the following way:

<u>dil_A</u> [<u>eros_A</u>]:

i) Centre the structural element $\hat{B}$ at the picture element z_i : $\hat{B} \dashrightarrow \hat{B} \oplus \{z_i\} =: \hat{B}_{z_i}$

ii) At every point x_j, which is covered by $\hat{B}_{z_i}$, find the minimum [maximum] grey value out of $m_A(x_j)$ and $m_B(x_j - z_i)$ [$1 - m_B(x_j - z_i)$].

iii) Replace $m_A(z_i)$ by the largest [smallest] of the minimum [maximum] values from ii).

An example for erosion was carried out and printed by the image processing equipment ROBOTRON A 6471 (see Fig.2). Fig.2a shows a grey picture with 191 x 97 picture elements and with five grey levels 0, 0.25, 0.5, 0.75, 1.0 . Fig.2b presents eros A eroded by a structural element B with 3 x 3 elements, where the centre point is black, the four corner points have grey level 0.25 and the remaining four points 0.75 .

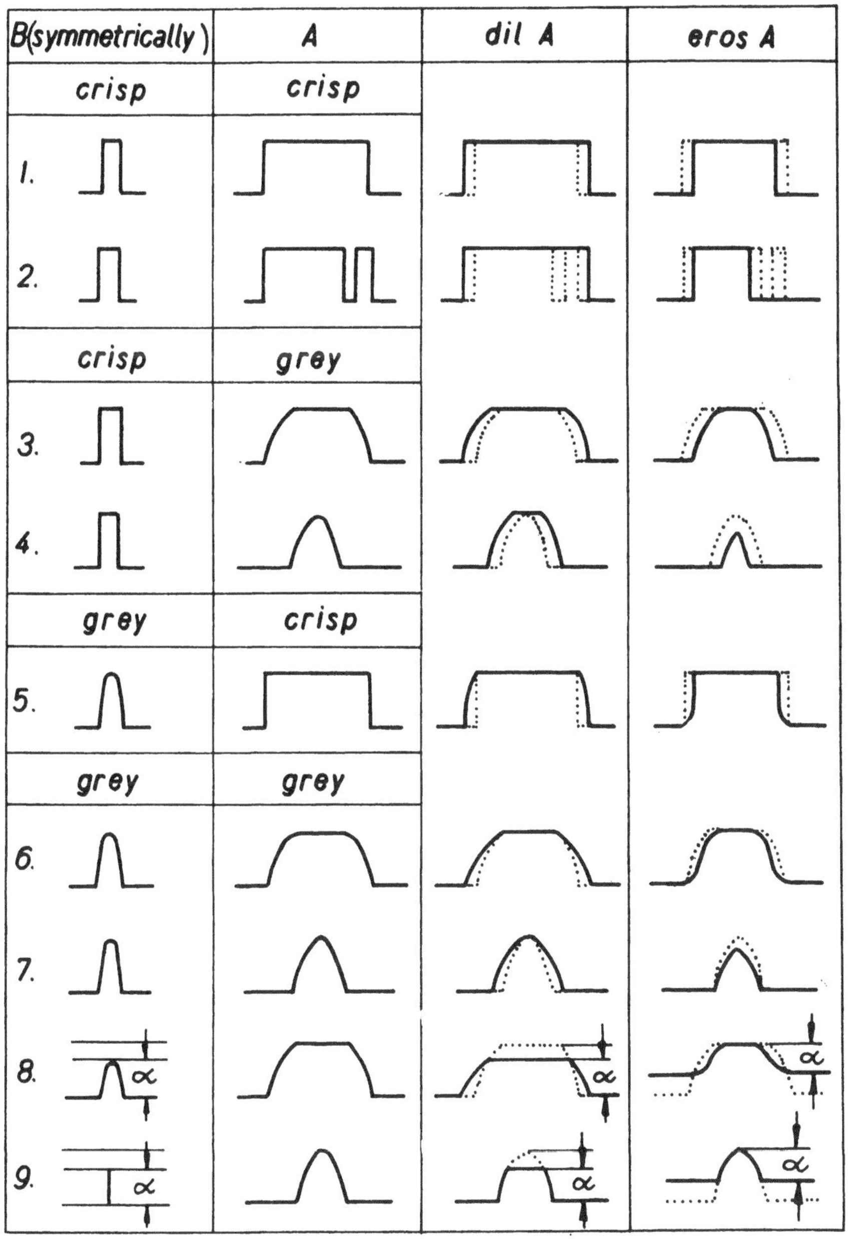

Fig.1: Some examples for dilatation and erosion of one-dimensional fuzzy sets

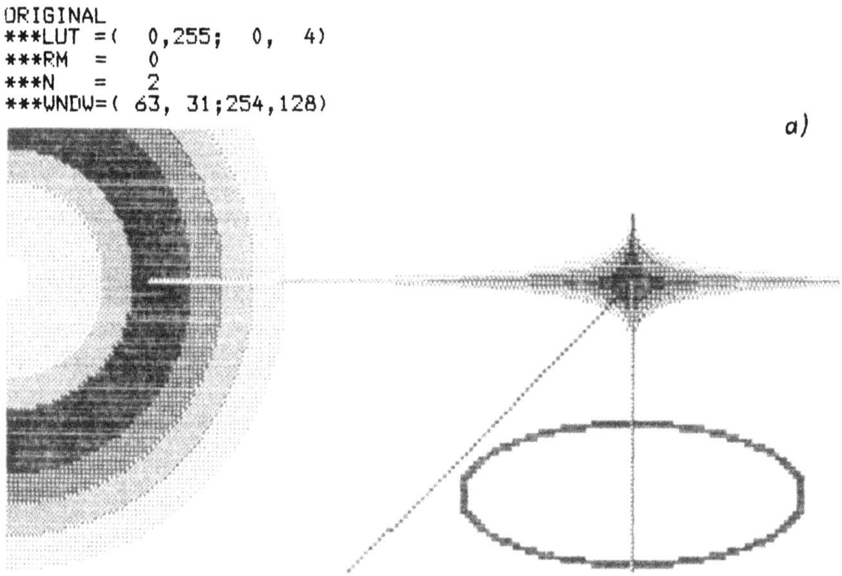

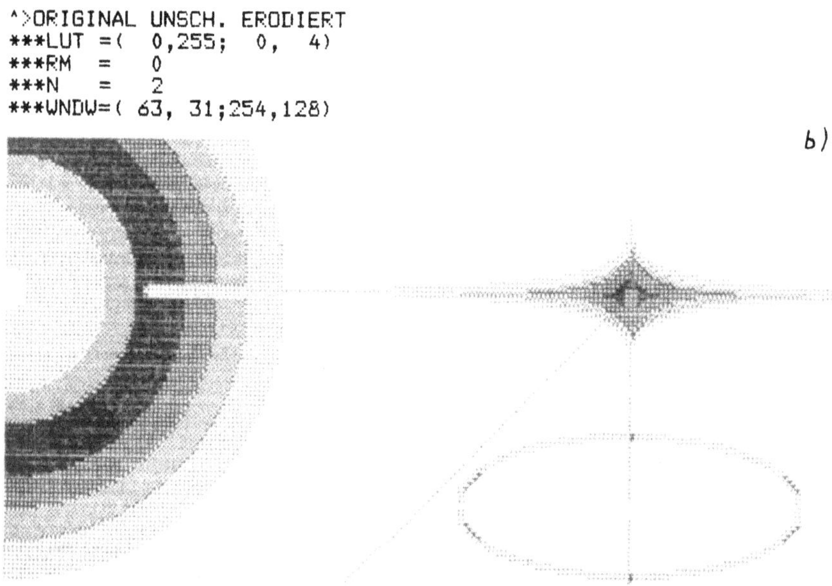

Fig.2: a) Grey A

b) Grey A eroded by grey B

Surely, image processing techniques with grey structural elements can be used as soft variants of the associated binary techniques. But here we will discuss another application. We conjecture that fuzzy image processing can help to develop practical algorithms to solve statistical problems for random fuzzy sets, possibly in a similar manner as binary image processing is an indispensible tool in statistics for (crisp) random sets. Let us go into some detail, e.g., for estimation of the covariance of a random f.s.

Let Y be a d-dimensional random f.s. with m.f. m_Y (as defined, e.g., in Puri/Ralescu [11]) and $Y_h = Y \oplus \{h\}$; $h \in \mathbf{R}^d$.

Then

$$C(x,x+h) = E \min(m_Y(x), m_{Y_h}(x)) \tag{2.4}$$

can be considered as the covariance function of the random f.s. Y . Note that for crisp Y with m.f. $\mathbf{1}_Y$ (2.4) reduces to

$$C(x,x+h) = E\, \mathbf{1}_Y(x)\, \mathbf{1}_{Y_h}(x) = P(x \in Y \cap Y_h) \tag{2.5}$$

which coincides with the usual definition of the covariance for random sets (see, e.g., Stoyan / Kendall / Mecke [13]). Using (1.3) formula (2.4) writes

$$C(x,x+h) = E\, m_{Y \cap Y_h}(x) \tag{2.6}$$

A straightforward consideration yields, also for fuzzy sets

$$Y \cap Y_h = Y \ominus \hat{B} \;;\; \hat{B} = \{0,h\} \tag{2.7}$$

Assume that Y is a stationary random f.s., i.e. $C(x,x+h) = C(h)$. Then $C(h)$ can be estimated by the mean grey value of the realised picture Y eroded by $\{0,h\}$, taken over a suitable grid $x_1,\dots,x_n$ of picture elements, i.e.

$$\hat{C}(h) = n^{-1} \sum_{i=1}^{n} m_{Y \ominus \{0,-h\}}(x_i) \tag{2.8}$$

This is a direct generalization of the usual method for estimation of the covariance for crisp random sets (see Stoyan / Kendall / Mecke [13]).

3. A fuzzy version of a shape factor

A central problem of particle shape analysis consists in the characterization of the shape of any natural or produced particle by a small number of parameters, which are to be evaluated from a two-dimensional picture of the particle, e.g. from its projection.

There are , essentially, two approaches to come to shape characterizations. The first approach suggests some transformation to get the contour sharp-edged and then an analysis of this contour by a Fourier series approximation. The Fourier coefficients or given functions of them are used as shape characteristics (cf. e.g. Beddow [4]; and for

critical remarks e.g. Bandemer / Albrecht / Kraut [1]).
The alternative approach consists in an, even subjective, valuation of some shape parameters such as roundness, sphericity and the like. Beddow [5] distinguishes, in his context, between long range pattern of relationship among the coordinates of the surface points (shape in the narrow sense) and short range pattern (defining the texture of the surface). Following a fuzzy theory line he proposes to value the shape (in the narrow sense) in specifying a value of sympathy with respect to each number of a given system of planar standard shapes, e.g. regular polygons with increasing number of vertices and changing relation between breadth and height, which he calls a 'fuzzy comparator'. The difficulty lies here in the specification of the values.
Our starting point is that the contour obtained by technical equipment is a grey-tone picture, e.g. reflecting the spatial nature of the particle, and that a sharp-edging transformation can change our impression of the shape essentially, and, moreover, can inspire a rather euphoric meaning with respect to the preciseness of the procedure performed. A workshop solution for a computer-aided particle shape description procedure, some sort of 'automatic fuzzy comparator', was presented by Bandemer / Kraut [3]. Because this procedure seems, at present still, to be somewhat too laborious, we consider a shape factor, characterizing roundness, for grey-tone pictures interpreted as fuzzy regions.
A usual shape factor for a crisp region G in $\mathbf{R}^2$ is the quotient formed by the perimeter of the contour $p = \text{card}_1 \partial G$ and of the perimeter of the circle covering the same area, $a = \text{card}_2 G$. For convenience we denote this shape factor (cf. Lueger [9]) by s^{-1}:

$$s^{-1}(G) = \text{card}_1 \partial G \,/\, 2\,(\pi)^{1/2}(\text{card}_2 G)^{1/2} = p \,/\, 2\,(\pi)^{1/2} a^{1/2} \qquad (3.1)$$

This factor equals one if G is a disk itself and s^{-1} is larger than one in any other case. Hence we can use the inverse value s as a degree of similarity with the disk, a special meaning of roundness, in the sense of fuzzy set theory. For example we have s(square) = 0.886; s(rectangle with a/b=0.5) = 0.835; s(rectangle with a/b=0.25) = 0.709; s(equilateral triangle) = 0.777 .
In analogy, in $\mathbf{R}^3$, we can introduce, as a degree of similarity with the ball, a special meaning of sphericity, the quotient of the area of the surface, $\text{card}_2 \partial G$, and of the area of the sphere covering the same volume, resulting in the shape factor

$$s^{-1}(G) = \text{card}_2 \partial G \,/\, (36\pi)^{1/3}(\text{card}_3 G)^{2/3}$$

In contrast with (3.1) this formula requires for its application in practice the determination of area and volume by physical methods.
Now, we assume that $P = \text{card}_1\, G$ and $A = \text{card}_2 G$ are given as f.s. s by their corresponding m.f.'s, say m_P and m_A. How to find such functions we will reflect upon later on.

Using Zadeh's extension principle we obtain the <u>fuzzy roundness</u> S by its m.f.

$$m_S(s) = \sup_{p,a\,:\,s=2(\pi a)^{1/2}/p} \min \{ m_A(a) , m_P(p) \} \qquad (3.2)$$

If P and A are given as fuzzy numbers, perhaps in LR-representation, we can write symbolically

$$S = 2(\pi)^{1/2} \odot A^{1/2} \odot P^{-1} \qquad (3.3)$$

and compute step by step

$$m_{A^{1/2}}(a) = \sup_{z\,:\,a=z^{1/2}} m_A(z) = m_A(a^2) \qquad (3.4)$$

$$m_{P^{-1}}(u) = \sup_{z\,:\,u=z^{-1}} m_P(z) = m_P(u^{-1}) \qquad (3.5)$$

using 'fast operations' (see Dubois / Prade [6]) to perform (3.3).
A similar calculus can be established for fuzzy sphericity.
Now we turn to the specification of m_A and m_P. The assumption that G is well-shaped (see (1.14) and (1.15)) is little restrictive for practical application. When computing $a = \mathrm{card}_2 G_\alpha$ for $\alpha \in (0,1]$ it can happen that only a few different values are met, e.g. if G looks like a staircase. Because of the monotonicity of G_α with respect to set inclusion we can define m_A for all $a \in \mathbf{R}^1$ by

$$m_A(a) = \sup \{\alpha : \mathrm{card}_2 G_\alpha \geq a \} . \qquad (3.6)$$

In contrast with $\mathrm{card}_2 G_\alpha$, the perimeter $\mathrm{card}_1 \partial G_\alpha$ does <u>not</u> show monotonicity for increasing α, in general. An analogous specification of m_P would favour the perimeter values for α's near one, which may lead to useless statements with respect to the shape factor. To guarantee monotonicity of $\mathrm{card}_1 \partial G_\alpha$ we can assume that G is <u>convex</u> (see (1.11)). In this case

$$m_P(p) = \sup \{\alpha : \mathrm{card}_1 \partial G_\alpha \geq p \} \qquad (3.7)$$

is a useful definition of the fuzzy perimeter P of G.
The value $m_A(a)$ can be interpreted as the degree to which 'the area of G is at least a and $m_P(p)$ evalues the corresponding statement with respect to the perimeter.
But even if both, m_A and m_P, are monotonic, the m.f. m_S can have several modal values. It is a problem of further investigation to find conditions on G to get m_S unimodal.
The value $m_G(x)$ gives the degree to which x belongs to G. If $m_G(x) < 1$ then we may conclude that x belongs to the fuzzy border ∂G of G, but the degree of this membership is not yet specified. We can avoid a strict decision if we introduce a <u>fuzzy set of second type</u>, i.e. for every value, say $\gamma \in [0,1]$, of membership we define a fuzzy set B, the fuzzy <u>γ-border of G</u>, by

$$m_B(x ; \gamma) = \min \{ m_G(x) / \gamma , ((1 - m_G(x)) / (1 - \gamma) \} . \qquad (3.8)$$

To return to an ordinary fuzzy set we can specify some γ, crisp or fuzzy, to express our prior knowledge or feeling, for which γ the border will make sense in the practical problem. In a context of particle shape description (see Bandemer / Kraut [3]) we took $\gamma = 0.5$ in favour of the cross-over-line $m_G(x) = 0.5$.

There is also a connection to the notion of a twofold fuzzy set F (see Dubois / Prade [7]), i.e. an ordered pair (I,C) with $I \subseteq \text{core}\, C$, where I is called the interior of F and C the closure of F. The boundary B of F is then defined by $B = C \cap \tilde{I}$ with

$$m_B(x) = \min \{ m_C(x) , 1 - m_I(x) \}. \tag{3.9}$$

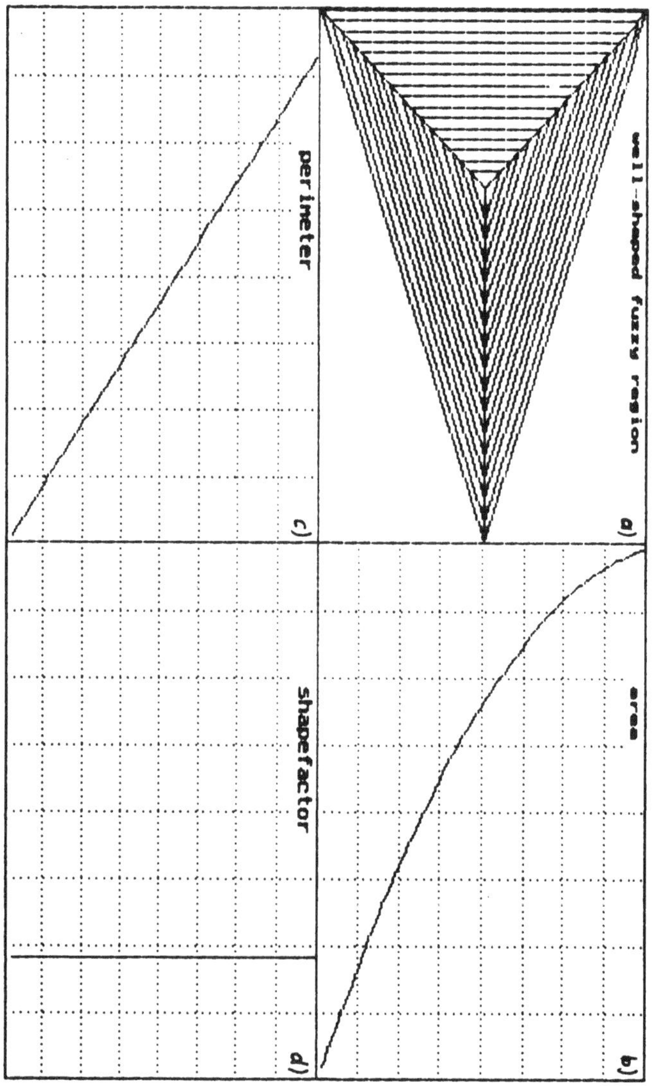

Fig.3: a) Fuzzy triangle; and the m.f.'s of its
b) area; c) perimeter and d) shape factor

(For interpretation within the framework of fuzzy set theory and more details see the paper cited above.)

In Fig.3 and Fig.4 two simple examples are provided to demonstrate the nature of computations and results. Fig.3 shows the m.f.'s of the fuzzy area, perimeter and shapefactor of a fuzzy triangle. In Fig.3a the fuzzy triangle is given by its α-cuts. Because these cuts have all the same shape, the resulting shape factor is crisp. Then Fig.4 shows the corresponding items for a fuzzy quadrangle with α-cuts of different shape. Here the shape factor is a fuzzy (non-crisp) one.

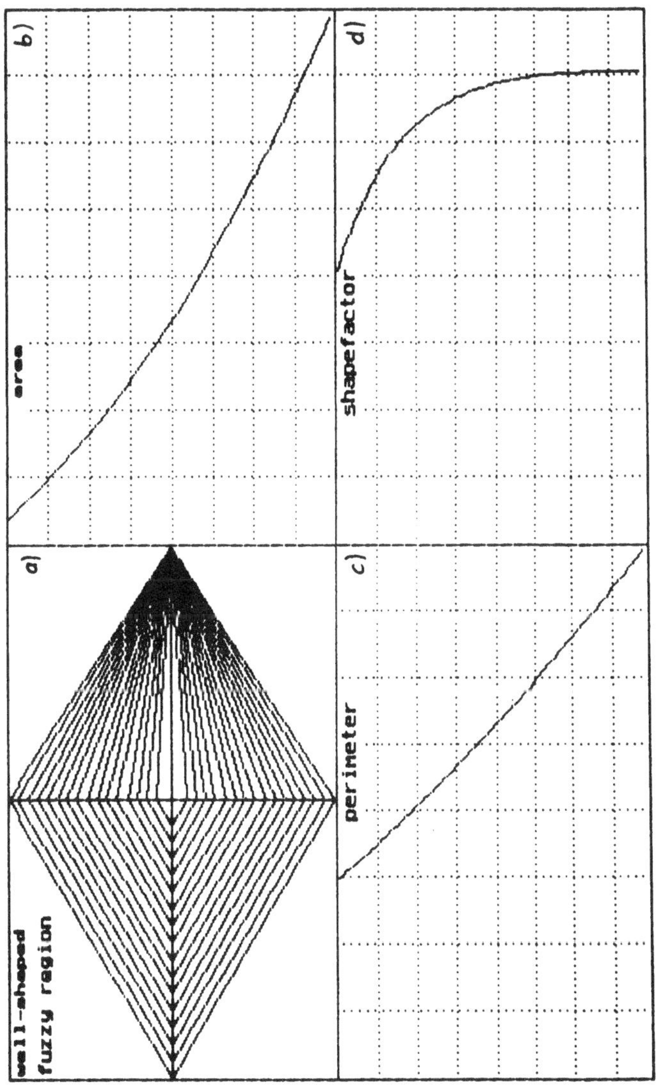

Fig.4: a) Fuzzy quadrangle; and the m.f. s of its b) area; c) perimeter and d) shape factor

References

[1] Bandemer, H., M.Albrecht, and A.Kraut: On using Fourier series in characterizing particle shape.
Part. Charact. 2 (1985), 98 - 103

[2] Bandemer, H. and S.Gottwald: Einführung in Fuzzy-Methoden. Theorie und Anwendung unscharfer Mengen.
Akademie-Verlag Berlin 1989

[3] Bandemer, H. and A.Kraut: On a fuzzy-theory-based computer-aided particle shape description.
Fuzzy Sets and Systems 27 (1988), 105 - 113

[4] Beddow, J.K. (Ed.): Particle Characterization in Technology, Vol.2: Morphological Analysis.
CRC Press, Boca, Raton, Fl., 1984

[5] Beddow, J.K., A.F.Vetter, and K.Simon: Powder Metallurgy Review 9, Part 2: Particle Shape Analysis.
Powder Metallurgy International 8(3), (1976), 107 - 109

[6] Dubois, D. and H.Prade: Fuzzy Sets and Systems: Theory and Applications.
Academic Press, New York 1980

[7] Dubois, D. and H.Prade: Twofold Fuzzy Sets: An approach to the representation of sets with fuzzy boundaries based on possibility and necessity measures.
Fuzzy Mathematics 3 (1983), 4, 53 - 76, Wuhan

[8] Goetcherian, V.: From binary to grey-tone image processing using fuzzy logic concepts.
Pattern Recognition 12 (1980), 7 - 15

[9] Lueger Lexikon der Technik, Bd.16
Deutsche Verlagsanstalt Stuttgart 1970

[10] Näther, W. and A.Kraut: Grey-tone image processing with fuzzy structural elements.
to appear in Systems Analysis-Modelling-Simulation

[11] Puri, M. and D.Ralescu: Fuzzy Random Variables.
Journ.Math.Anal.Appl. 114 (1986), 409 - 422

[12] Serra, J.: Image Analysis and Mathematical Morphology.
Academic Press, New York 1982

[13] Stoyan, D., W.D.Kendall, and J.Mecke: Stochastic Geometry and Its Application.
Akademie Verlag / John Wiley & Sons, 1987

[14] Zadeh, L.A.: Fuzzy Sets.
Information and Control 8 (1965), 338 - 353

[15] Zadeh, L.A.: Fuzzy Sets and Applications. Selected Papers.
(R.R.Yager et.al.Eds.) Wiley, New York 1987

Polyhedral Object Recognition by Hough Space Analysis

T.G. Stahs and F.M. Wahl

Institute for Robotics and Computer Control
Technical University of Braunschweig, FRG

Abstract: It is known, that recognition of ployhedral objects in single 2D images of 3D scenes can be performed advantageously by analysing cluster patterns in Hough space. The efficiency of this technique is based to a great deal on the extraction of structural and geometrical object features only available under orthographic projection. In this paper we propose a technique to reconstruct an orthographic view of a perspective image of a polyhedral scene by detecting vanishing points and subsequently orthogonalizing Hough nets in Hough space. Beyond this, the knowledge of vanishing points allows the extraction of 3D information from the scene resulting in a considerable reduction of the number of candidate models for the recognition procedure.

1. Introduction

In numerous publications the Hough transform (HT) is merely regarded as a technique to estimate parameters of analytically or non-analytically given curves or surfaces. Therefore most efforts in the Hough arena concentrated on the HT itself rather than on an analysis of cluster patterns in the parameter or Hough space (HS) [WAHL88a]. Nevertheless, some simple structural relationships between clusters in HS - esp. collinear arrangements of cluster locations (in short: collinearities) - allow extensive reasoning about the 2D-image and the underlying 3D-scene.

Based upon these relationships we propose a two-stage technique for the recognition of three-dimensional polyhedral objects from single grayscale images of a scene. The starting point in our further discussion is line information obtained from the image by the HT. The expense for this transformation can be reduced considerably by applying a special edge detector extracting gradients as well as gradient directions (see e.g. [BURO79]). This reduces the HT to a simple histogramming technique in the 2D (slope, intercept)-parameter space. Subsequently, similar as in our approaches described e.g. in [WAHL85, ENGE88], topological and geometrical features of the scene are extracted by analysing the cluster patterns in HS. This first stage yields an attributed graph representation of the observed object, which is compared to similar representations of wire frame model objects by means of an attributed subgraph isomorphism detection algorithm. The second stage then verifies the resulting topological matches geometrically using model-point/image-line correspondences. The camera model calculated for this purpose enables the determination of position and orientation parameters of recognized objects relative to a camera system.

Our previous work [ENGE88, WAHL88b, WAHL88c] described HS analysis techniques under an orthographic projection assumption. In this paper we will concentrate on the HS analysis with special regard to perspective projections. On efficient realization of the HT as well as on determination of position and orientation parameters more details may be found in [NOEL88] and [WAHL88b].

2. Polyhedral object recognition under orthographic projection

For simplicity we assume the HS to be produced by the original HT applied to a binary gradient image of a 3D-scene using a (slope, intercept)-parameter space [HOUG62]. The unboundedness of the HS is assumed to be resolved by the twin HS representation proposed in [ROSE69]. In this section we further assume that the images from which the HSs are obtained show objects from non-degenerate viewpoints with an infinite focal distance. A non-degenerate viewpoint specifies, that no edges are accidentally aligned in the image and that any threefold or higher-degree intersection of the infinite extensions of these edges corresponds to an actual object vertex. Regarding these preconditions the following relationships between HS, image space (IS) and the 3D scene (3DS) hold:

- A cluster in HS corresponds to a line or a set of collinear lines in IS; this corresponds to one or more edges in the 3DS.
- Vertically aligned clusters in HS correspond to parallel lines in IS and to parallel edges in 3DS.
- n collinear clusters in HS ($n \geq 3$) correspond to n lines intersecting each other at one location in IS or to a vertex of order $\geq n$ in 3DS.
- n collinearities sharing one cluster correspond to one line intersected by at least two lines at n different locations in IS or to n collinear vertices in 3DS; the edge corresponding to the common cluster is called super-edge for $n \geq 3$.

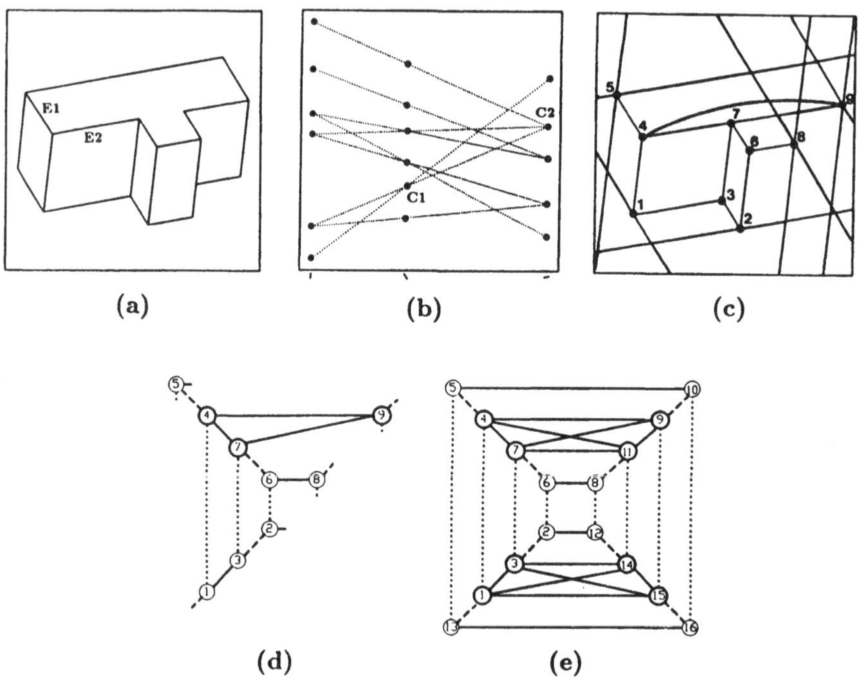

Fig. 1: Polyhedral object recognition by Hough space analysis and subgraph isomorphism detection.

Clusters and collinearities form so-called Hough nets (HNs). Fig. 1b shows the HN for the object in Fig. 1a. A closer look reveals three groups of vertically alligned clusters called slope-classes. For clusters participating in two collinearities - e.g. C1 - it is trivial to calculate length of the corresponding image edge E1. This length is of course viewpoint-dependent; it can be used to calculate and compare length ratios in the object identification stage. Super-edges are decomposed into $n!/[2(n-2)!]$ normal image edges. Fig. 1c shows a representation of the object after transforming clusters from HS back into IS taking into account all possible length calculations and a super-edge decomposition for C2.

The results of HS analysis can be represented as attributed graphs in which each node represents a vertex and each arc represents an edge of the observed object. A boolean attribute for each node indicates the participation of the corresponding vertex in a super edge whereas the arcs are specified by the slope class and the length of their image edge (if computable). Fig. 1d shows the graph representation of the object in Fig. 1a.

The object recognition is realized by matching this image graph against a set of model graphs using an attributed subgraph isomorphism detection algorithm. The model graphs are generated form non degenerate 2D wire frame representations of the 3D model objects; after an analytical HT the HS analysis mentioned above is applicable here too. The subgraph detection algorithm is an extension of Ullmann's algorithm for matching non-attributed graphs [ULLM76]. The integration of graph attributes and experimental results are described in detail in [ENGE88]. However, it is important to see, that a substantial acceleration of the subgraph isomorphism detection by means of graph attributes requires a certain amount of parallel structures in the objects of the model world - a requirement fulfilled in many possible applications.

3. Hough space under perspective projection

A limiting restriction for the analysis described above is the assumption of orthographic views. Hence it is essential to make this technique applicable for perspective projections too. The main problem arising here is the fact, that parallel 3D object edges generally don't produce parallel image edges. Thus, most of the geometrical object features under orthographic projection, like slope classes or length ratios, are no longer available in the perspective case. To understand how to overcome these difficulties, it is important to point out some properties of HSs of perspective views.

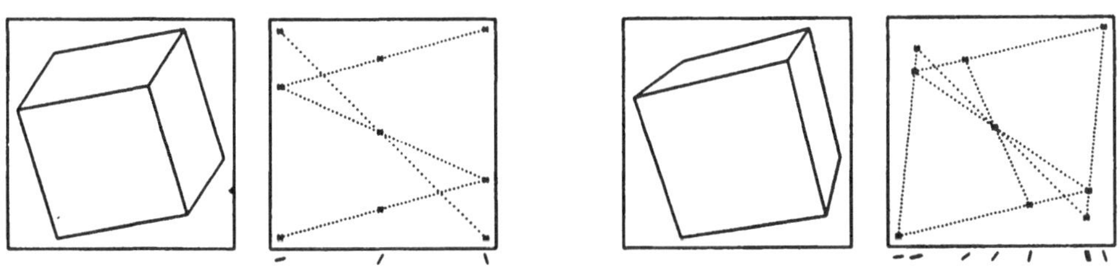

(a) (b)

Fig. 2: Hough space under othographic (a) and perspective (b) projection.

Fig. 2 shows, that slope classes turn into 'normal' collinearities under perspective projection. These collinearities correspond to the vanishing points of the images of parallel 3D object edges. Thus, under our non-degenerate viewpoint assumption each collinearity in the HS indicates either an object vertex or a vanishing point in IS.

A variation of the focal distance shows another important property of perspective HSs: vertices as well as vanishing points move along straight lines L_i through the principal point H in IS (Fig. 3a). In HS H and L_i correspond to a straight line H' and to points L_i' on this line respectively. They are invariant with respect to a variation of the focal distance. On the other hand, since the vertices and the vanishing points move along the invariant lines L_i in IS, the fixed points L_i' are part of the corresponding collinearities. Fig. 3b shows that a variation of the focal distance causes the collinearities to rotate around these fixed points in HS. In course of this rotation the participating clusters move along straight lines in HS (e.g. line C in Fig. 3b). In IS these lines correspond to the intersection of infinite extensions of 3D object edges with the image plane.

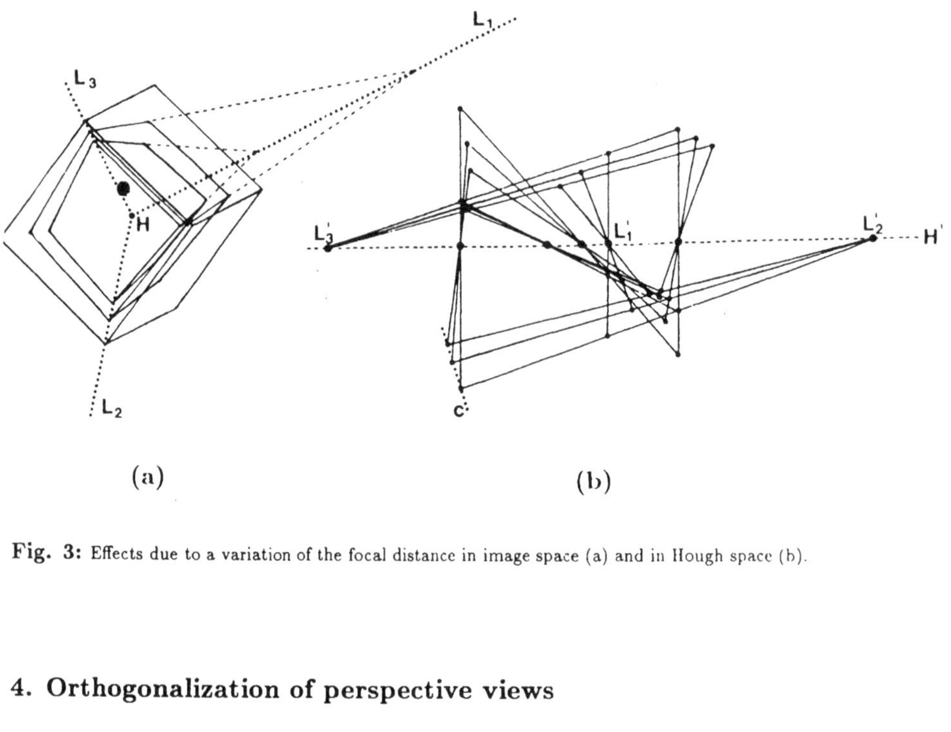

Fig. 3: Effects due to a variation of the focal distance in image space (a) and in Hough space (b).

4. Orthogonalization of perspective views

Based upon the properties of perspective HSs we will now show how to reconstruct an orthographic view of a perspective image of a polyhedral scene. The main idea is to 'orthogonalize' the HS by rotating the vanishing point collinearities in an upright position maintaining all structural relationships between clusters and collinearities in the HN. An essential problem in this context is the distinction between vertex and vanishing point collinearities in HS. This question will be considered in more detail in section 5. Here we want to illustrate our orthogonalization procedure by means of a simple example.

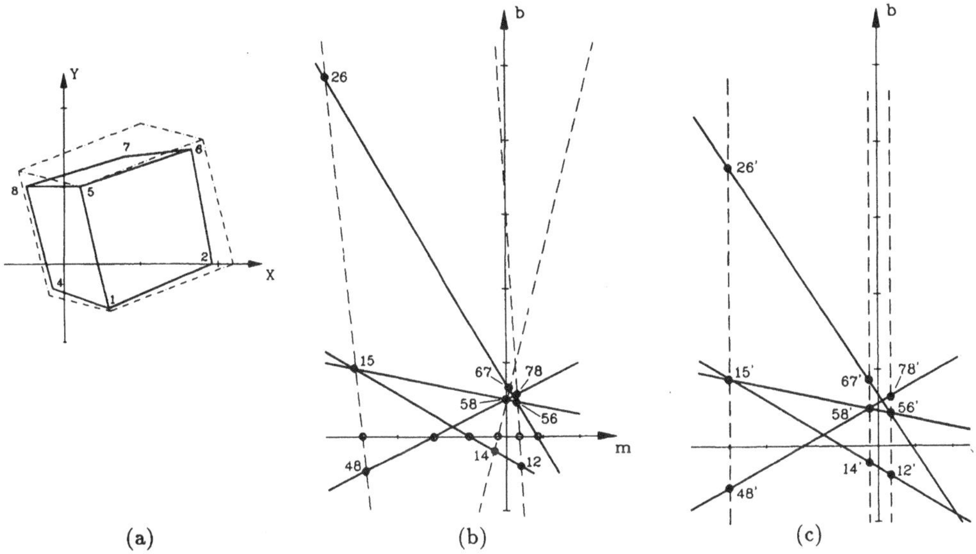

Fig. 4: Orthogonalization of the Hough net of a cube.

Fig. 4a shows the perspective view of a cube, Fig. 4b the corresponding HN. For simplicity we assume, that the optical axis of the camera intersects the image plane in the origin of the x,y-coordinate system. If the vanishing point collinearities are known (dashed lines in Fig. 4b), we can put them upright by rotating them around their fixed points on the m-axis. The new position of an arbitrary start cluster, e.g. 12†, is determined by moving it parallel to the m-axis onto the orthogonalized vanishing point collinearity. It's new position and the fixed point of the collinearity corresponding to vertex 1 define the new position of this vertex collinearity and thus the new position of the clusters 14 and 15. A recursive procedure determines the new positions of the remaining clusters and collinearities in the HN (Fig. 4c). The dot-dashed lines in Fig. 4a show the reconstructed orthographic image of the cube.

Note, that we don't move the start cluster along the line defined by a variation of the focal distance (e.g. line C in Fig.3b). In fact, we can't reconstruct this line from a single perspective view, but it is important to see, that the selection of the start cluster and the direction of it's displacement merely affect the scale factor for the reconstructed orthographic image.

In some cases the application of the recursive procedure described above does not involve all clusters in the HN. On the one hand, this is true for clusters which participate in at most one collinearity. In general, they represent only little structural information; the orthogonalization procedure merely has to retain them on their collinearities. On the other hand it is possible that the structural connection between two groups of clusters isn't tight enough. In this case the recursion has to be restarted with a new start cluster. This re-application results in several differently reconstructed orthographic subimages. Each of these subimages is scaled with a different scale factor. This is important especially for the calculation and comparison of lenght ratios in the graph matching procedure.

† cluster ij corresponds to the edge between the vertices i and j in IS.

5. Detection of vanishing points

A problem already mentioned in section 4 is the distinction between vertex and vanishing point collinearities. Under our non degenerate viewpoint assumption collinearities corresponding to points outside the image border can be regarded as vanishing points. If all clusters are participating in such collinearities the orthogonalization procedure can be started immediately. Otherwise vanishing point hypotheses are generated for the remaining collinearities. The number of possible hypotheses can be reduced considerably by regarding the fact, that a cluster can participate at most in one vanishing point collinearity. The hypotheses are accepted or rejected with the aid of 3D information from the model world (see below). Since the detection of vanishing points is crucial for the orthogonalization procedure, we also generate and test vanishing point hypotheses for pairs of clusters which do not participate in a vanishing point collinearity up to now. For the remaining clusters it is impossible to determine vanishing points.

The verification of the vanishing point hypotheses with the aid of 3D information from the model world is based on the following consideration (Fig. 5a):

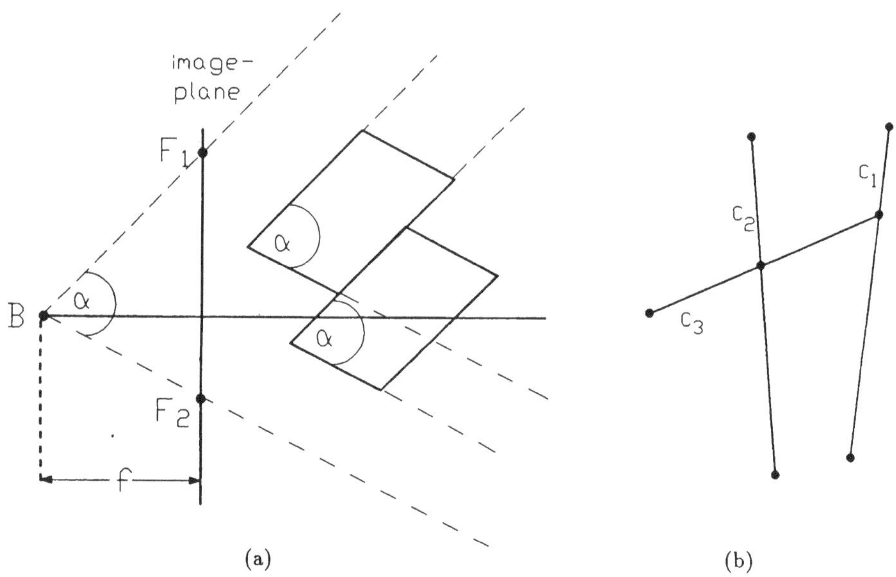

Fig. 5: (a) Relationship between vanishing points in the image plane and object vertices in the scene. (b) Configuration of collinearities in Hough space for testing vanishing point hypotheses.

Let F_1 and F_2 be the vanishing points for two classes of parallel object edges. In connection with the focal distance f they determine an angle α. This angle reappears at every intersection of two edges of these classes in 3DS.

If F_1 and F_2 correspond to collinearities C_1 and C_2 respectively, the main idea for the verification of a vanishing point hypothesis F_1 is as follows:
Let C_2 be an already known vanishing point collinearity and C_3 a collinearity connecting C_1 and C_2

(Fig. 5b). If the hypothesis F_1 is correct, there must exist a vertex in the 3D model world with two edges intersecting each other under the angle α. If such a vertex doesn't exist, the hypothesis has to be rejected; else it is retained. In unfavourable constellations the last decision might be wrong. But the probability of incorrect decisions can be minimized by collecting as much evidence as possible before accepting the hypothesis (i.e. trying to verify it by using other already known vanishing point collinearities). However, incorrect decisions are detected at latest in the geometrical match.

The search for vanishing points not only enables the orthogonalization of perspective views but also allows a substantial acceleration of the graph matching process. Since the acceptance of a vanishing point hypothesis is based on the existence of certain vertices in the model world, only the models containing these vertices need to be considered in the graph matching process. Moreover, the acceptance of a vanishing point hypothesis considerably reduces the possible matches between image vertices and the vertices of a candidate model. Therefore it is adventageous to apply the verification procedure even for vanishing points outside the image border.

6. Conclusion

Our investigations indicate, that polyhedral object recognition by HS analysis is a feasible technique even for perspective views. Geometrical object features variant under perspective projection are regained by reconstructing an orthographic view of the scene. This is achieved by detecting vanishing points and orthogonalizing HNs in HS. Thus, an efficient attributed graph matching algorithm realizing the polyhedral object recognition for orthographic views can be employed in the perspective case too. In addition to this, the number of candidate models for the graph matching procedure can be reduced considerably by taking into account 3D edge direction information about the scene extracted from HS.

Our current research concentrates on evaluation of the robustness of our technique with respect to fuzzy HSs due to noisy images and imperfect low-level image processing - a problem closely related to our non-degenerate viewpoint assumption. An investigation of absence or fusion of clusters as well as the effects of accidential collinearities on the performance of the recognition process is important esp. for processing real world scenes. In this context it is worth mentioning, that we expect our recognition procedure to be applicable for 'polyheder-like' objects too, i.e. objects with enough straight line edges to be discriminated on this basis. Many workpieces in industrial environments fulfill this requirement.

Acknowledgement

The authors wish to thank H. Mangold for the implementation and evaluation of the proposed technique.

References

BURO79 M. Burow, F.M. Wahl "Eine verbesserte Version des Kantendetektionsverfahrens nach Mero/Vassy", in Foith J.P. (ed.): "Angewandte Szenenanalyse", Informatik Fachberichte 20, Springer (1979); also described in: F. Wahl, "Digital Image Signal Processing", pp. 141, Artech House (1987).

ENGE88 J.R. Engelbrecht, F.M. Wahl "Polyhedral Object Recognition Using Hough-Space Features", *Pattern Recognition*, **21**, No. 2, Pergamon Press (1988).

HOUG62 P.V.C. Hough, A. Arbor "Method and Means for Recognizing Complex Patterns", US Patent No. 3'069'654 (1962).

NOEL88 M. Nölle "Lage- und Orientierungsbestimmung von 3d Objekten aus 2d Linienparametern", Institute of Robotics and Computer Control, Technical University of Braunschweig, FRG (1988).

ROSE69 A. Rosenfeld chapter 8.4, pp.151: "Coordinate Conversion", in A. Rosenfeld "Picture Processing by Computer", *Academic Press* (1969).

ULLM76 J.R. Ullmann "An Algorithm for subgraph Isomorphism", *J. ACM*,**15**, No. 1 (1976).

WAHL85 F.M. Wahl "Method for Identifying Three-Dimensional Objects Using Two-Dimensional Images", European Patent No. 0'205'628 (1985); US Patent No. 4'731'860 (1988).

WAHL88a F. Wahl "A Survey of Hough Transform Techniques", Tech. Report 5-88-1, Institute of Robotics and Computer Control, Technical University of Braunschweig, FRG (1988).

WAHL88b "Analysing Hough Nets for Recognition of Polyheder-Like Objects", Proc. of 9th Conf. on Pattern Recognition, Rome, Italy, IEEE press (1988).

WAHL88c F. Wahl "Deriving features from Hough space for object recognition and configuration estimation", to appear in: J.C. Simon (ed.): "From the pixels to the features", *North Holland* (1988).

QUADTREES AND THE HAUSDORFF DIMENSION OF PICTURES

LUDWIG STAIGER[1)]

Many important spatial patterns of nature are irregular in such a degree that "smooth" models are not suitable for describing their form. As an appropriate means for solving this problem it turned out to consider so-called fractals [5]. One domain of application of fractal models is the estimation of the Hausdorff dimension of a finely structured surface (or picture). Quadtrees [2,3] have recently become known as a convenient means of encoding 2-dimensional (black-and-white) pictures because they enable large picture fragments of the same colour to be represented by a single node while retaining spatial information in a form suitable for processing.

In this paper we show that the quadtree can also be utilized when calculating the Hausdorff dimension of a picture (a surface). For theoretical reasons we use an infinite quadtree representing a picture in the unit square $[0,1]\times[0,1]$ in the following manner (the connection to finite quadtrees being explained below):

The quadtree is a rooted directed tree generated by the recursive subdivision of the picture into quadrants. The root represents the whole picture, and each node has four sons, one for each quadrant of the subpicture belonging to that node.

Our approach is based on the following fact: If a subquadrant is not relevant for the object under observation, e.g. if it belongs entirely to the background of the picture, then the node representing this quadrant is labelled $\emptyset$ (empty set) and is removed from the qudtree together with all its successors. Furthermore, any node without successors is also removed from the tree. (If the picture consists entirely of background this procedure results in an empty quadtree; in the sequel we shall disregard this trivial case.)

The peculiarity of our approach is that we do not care what the picture really is, but only how it behaves when we successively subdivide it, since we assume the Hausdorff dimension of a picture P in the unit square to be the dimension of the underlying support $\mathrm{supp}P := \{(x,y): 0 \leq x,y \leq 1$ and (x,y) belongs to $P\}$ and our subdivision process successively removes quadrants not belonging to this support. Thus, if we assume the support to be a closed (in the usual topology) subset of the unit square, our approach yields the real Hausdorff dimension of the picture.

In practice, however, one is usually confined to quadtrees of $2^n\times 2^n$ pictures, i.e. to finite quadtrees. In order to extend a finite quadtree to an infinite one similarities between the quadrants of the subdivisions are used to replace smaller subtrees by larger ones in case when their roots belong to similar quadrants. If this process is continued in a manner similar to the extension of a partial finite automaton to a fully defined (finite) automaton (cf. [9]) we arrive, finally, at an infinite quadtree in which the nodes belong to only finitely many similarity classes. This latter fact will be advantageously used below.

1) Karl-Weierstraß-Institut für Mathematik der AdW
Mohrenstr. 39 PF 1304 Berlin DDR-1086

In order to formalize our approach the nodes of the infinite quadtree are denoted by words in the alphabet $Q := \{ (0,0), (0,1), (1,0), (1,1) \}$ in the following way: The root is denoted by e (the empty word), and the son of the node $w \in Q^*$ in direction $a \in Q$ ((0,0) ≙ south-west, (0,1) ≙ north-west, (1,0) ≙ south-east, and (1,1) ≙ north-east) is denoted by wa . Now it is easily verified that the infinite expansion (sequence) $\zeta \in Q^\omega$ converges to the point $z = (x,y)$ whose coordinates x and y have the dyadic expansions $\xi,\eta \in \{0,1\}^\omega$ which are the respective projections of ζ .
Thus, instead of estimating the real Hausdorff dimension of a picture, we may follow the line of [6] (cf. also [8]) of estimating the Hausdorff dimension of the corresponding set of Q-expansions.[1)]

NOTATION

We consider the space X^ω of infinite sequences on a finite alphabet X of cardinality card X = r . By X^* we denote the set of finite strings (words) on X . For $w \in X^*$ and $b \in X^* \cup X^\omega$ let $w \cdot b$ be their concatenation. This concatenation product extends in an obvious way to subsets $W \subseteq X^*$ and $B \subseteq X^* \cup X^\omega$ [2)] Furthermore, $|w|$ is the length of the word $w \in X^*$, and $\underline{\ell}(W) := \inf \{ |w| : w \in W \}$ and $\ell(W) := \sup \{ |w| : w \in W \}$ denote the minimum and maximum length resp. of words in $W \subseteq X^*$.
For a set $B \subseteq X^* \cup X^\omega$ we call $B/w := \{ b : w \cdot b \in B \}$ the *state* of the set B derived by the word $w \in X^*$, and we call B *finite-state* provided $\{ B/w : w \in X^* \}$ is a finite set.
We consider X^ω as a metric space with the metric ρ defined by

$$\rho(\xi,\eta) := \inf \{ r^{-|w|} : \xi,\eta \in w \cdot X^\omega \} \quad . \tag{1}$$

It is well known (cf. [4]) that this space (X^ω,ρ) is homeomorphic to the Cantor discontinuum, hence compact. The open balls in this space are the sets of the form $w \cdot X^\omega$.
If we identify an infinite quadtree with the set of all of its infinite paths $\eta \in Q^\omega$ starting from the root then one easily verifies that the identity of states corresponds to the similarity of subquadrants in the picture represented by the quadtree. Moreover, it is also easily verified that this set of infinite paths is a closed subset of Q^ω. Thus the above mentioned extension of finite quadtrees into infinite ones leads to finite-state closed subsets of Q^ω.

COVERINGS AND HAUSDORFF DIMENSION

An r^{-n}-*covering* of a set $F \subseteq X^\omega$ is a family $(v \cdot X^\omega)_{v \in V}$ of balls satisfying the condition $\bigcup \{ v \cdot X^\omega : v \in V \} = V \cdot X^\omega \supseteq F$ and whose diameters diam $v \cdot X^\omega$ do not exceed r^{-n}, i.e. $\underline{\ell}(V) \geq n$. We refer to a family $(v \cdot X^\omega)_{v \in V}$ as a *minimal covering* of a set F if no proper subfamily of $(v \cdot X^\omega)_{v \in V}$ covers the set F . Since $v \cdot X^\omega \supset w \cdot X^\omega$ implies that v is a proper

1) If we do so we have transferred the two-dimensional problem into a one-dimensional. Thus, in the final evaluation of the Hausdorff dimension of a picture we have to take care of a scaling factor of 2.

2) For the sake of brevity, in what follows we shall write $w \cdot B$ instead of $\{w\} \cdot B$.

prefix of w, the index set V of a minimal covering is *prefix-free*, i.e. V does not contain words $w \neq v$ such that one is an initial part of the other. Moreover, from the equivalence $F/w = \emptyset$ iff $w \cdot X^{\omega} \cap F = \emptyset$ we obtain that the index set of a minimal covering of F does not contain words w with $F/w = \emptyset$, and since (X^{ω}, ρ) is a compact space, a minimal covering of a closed subset of X^{ω} is always finite.
In the sequel, we abbreviate the fact that V is the index set of a minimal covering of the set F by $V \perp F$.
In Section 14 of [1] the general definition of the Hausdorff dimension of a subset of a metric space is given. We recall here this definition for our purposes.
We define

$$L_{\alpha}(F;V) := \sum_{v \in V} (\operatorname{diam} v \cdot X^{\omega})^{\alpha} = \sum_{v \in V} r^{-\alpha \cdot |v|} \tag{2}$$

for $V \cdot X^{\omega} \supseteq F$ and $0 \le \alpha \le 1$. Then

$$L_{\alpha}(F) := \liminf_{n \to \infty} \{ L_{\alpha}(F;V) : V \cdot X^{\omega} \supseteq F \wedge \underline{\ell}(V) \ge n \} \tag{3}$$

is the α-dimensional outer measure of F. (In this definition we could likewise replace $V \cdot X^{\omega} \supseteq F$ by $V \perp F$.)
Now, consider $L_{\alpha}(F)$ for fixed F as a function of α. Then there is an $\alpha_0 \in [0,1]$ such that $L_{\alpha}(F) = \infty$ if $\alpha < \alpha_0$ and $L_{\alpha}(F) = 0$ if $\alpha > \alpha_0$. This "change-over" point α_0 is called the *Hausdorff dimension* dim F of the subset F of X^{ω}, i.e.

$$\dim F := \sup \{ \alpha : L_{\alpha}(F) = \infty \} = \inf \{ \alpha : L_{\alpha}(F) = 0 \} . \tag{4}$$

Next, we consider the case when we use as coverings of F only the coarsest (but minimal) r^{-n}-coverings. Their index sets are easily identified as $V_n(F) = \{ w : |w| = n \wedge F/w \neq \emptyset \}$. The function $s_F : \mathbb{N} \to \mathbb{N}$ defined as

$$s_F(n) := \operatorname{card} V_n(F) \tag{5}$$

is known as the structure function of the language (subset of X^*) $\{ w : F/w \neq \emptyset \}$.
Comparing with the definition of $L_{\alpha}(F)$ we get the inequalities

$$L_{\alpha}(F) \le \liminf_{n \to \infty} s_F(n) \cdot r^{-\alpha \cdot n} \tag{6}$$

and consequently

$$\dim F \le \liminf_{n \to \infty} n^{-1} \cdot \log_r s_F(n) . \tag{7}$$

In the case of subsets F of X^{ω} related to quadtrees we obtain even equality in Eq. (7)

Theorem 1. [6,8] If $F \subseteq X^{\omega}$ is finite-state and closed then

$$\dim F = \lim_{n \to \infty} n^{-1} \cdot \log_r s_F(n) . \tag{8}$$

The quantity on the right hand side of Eq. (7) or (8) is also known as the entropy of the subset F of X^{ω} [7]. In Chapter 9 of [4] and in [7] it is shown that the entropy of a finite-state subset can be calculated in a simple manner. Now, our Theorem 1 enables us to apply these results for the estimation of the Hausdorff dimension.
To this end define the adjacency matrix A_F of a finite-state subset $F \subseteq X^{\omega}$ having the set of states $\{F_1 = F, F_2, \ldots, F_n\}$ as the $n \times n$ matrix whose (i,j) entry is

$$a_{ij} := \operatorname{card} \{ x : x \in X \wedge F_i/x = F_j \} \tag{9}$$

Together with the result of Theorem 1 we obtain from Eq. (17) of [7] the following.

Lemma 2. If F is finite-state and closed then $\dim F = \log_r |\lambda|$ where λ is an eigenvalue of A_F of maximum modulus.

α-DIMENSIONAL MEASURE

Depending on the set F the function L_α may have at the point $\alpha = \dim F$ any value $0 \le L_\alpha(F) \le \infty$. Since the metric ρ defined in Eq. (1) does not coincide with the metric in the real space $\mathbb{R}^2$, our measure L_α for $\alpha = \dim F$ differs from the real measure. But as it was pointed out in Section 14 of [1] there are constants $c_1, c_2 > 0$ such that

$$c_1 \cdot L_\alpha \le L_\alpha^{\text{real}} \le c_2 \cdot L_\alpha \tag{10}$$

Thus the methods below allow us to determine whether the α-dimensional measure of a picture is zero, finite, or infinite.

We start with a technical result.

Lemma 3. Let $F \subseteq X^\omega$ be closed, $0 \le \alpha$ and $0 < c$. Then

(i) $\forall w \forall i (w \in X^* \wedge i \in \mathbb{N} \rightarrow s_{F/w}(i) \le c \cdot r^{\alpha \cdot i})$ implies that for all V such that $V \cdot X^\omega \supseteq F$ and $\ell(V) \le n$ the inequality $L_\alpha(F;V) \ge \frac{1}{c} \cdot s_F(n) \cdot r^{-\alpha \cdot n}$ holds, and

(ii) $\forall w \forall i (F/w \ne \emptyset \wedge i \in \mathbb{N} \rightarrow s_{F/w}(i) \ge c \cdot r^{\alpha \cdot i})$ implies that for all V such that $V \perp F$ and $\ell(V) \le n$ the inequality $L_\alpha(F;V) \le \frac{1}{c} \cdot s_F(n) \cdot r^{-\alpha \cdot n}$ holds.

Proof. We show only (ii), (i) being proved in the same way.

We have $L_\alpha(F;V) = \sum_{v \in V} r^{-\alpha \cdot |v|} = \sum_{v \in V} \frac{r^{\alpha(n - |v|)}}{s_{F/v}(n - |v|)} \cdot s_{F/v}(n - |v|) \cdot r^{-\alpha \cdot n}$

$\le \frac{1}{c} \cdot r^{-\alpha \cdot n} \cdot \sum_{v \in V} s_{F/v}(n - |v|)$ by the hypothesis. Since $V \perp F$ and $\ell(V) \le n$,

$\sum_{v \in V} s_{F/v}(n - |v|) = s_F(n)$ follows. Q.E.D.

We obtain a first result.

Theorem 4. If $F \subseteq X^\omega$ is nonempty, finite-state and closed, $\alpha = \dim F$, then $L_\alpha(F) > 0$.

Proof. According to Theorem 14 and Corollary 4 of [7] the set F contains a finite-state closed subset E such that $c_1 \cdot r^{\alpha \cdot n} \le s_{E/w}(n) \le c_2 \cdot r^{\alpha \cdot n}$ for some $c_1, c_2 > 0$ and every w such that $E/w \ne \emptyset$. Now the assertion follows from Lemma 3. Q.E.D.

Our theorem establishes that in any case a nonempty, finite-state closed subset $F \subseteq X^\omega$ has nonzero (dim F)-dimensional measure. Next we are going to investigate when this measure is finite.

To this end we define for $L_\alpha(F) > 0$

$$E := F \setminus \bigcup \{ F \cap w \cdot X^\omega : L_\alpha(F \cap w \cdot X^\omega) = 0 \} . \tag{11}$$

It is easily observed that

$$L_\alpha(E) = L_\alpha(F) \text{ and } \forall w (E/w \ne \emptyset \rightarrow L_\alpha(E \cap w \cdot X^\omega) \ne 0) \tag{12}$$

Moreover, we need the following.

Property 5. Let (a_n), (b_i) and (c_j) be families of nonnegative real numbers such that $b_i \le b \cdot r^{\alpha \cdot i}$, $c_j \le c \cdot r^{\gamma \cdot j}$ for some constants b, c, α and γ where $\alpha > \gamma > 0$, and let

$$a_n = \sum_{i=0}^{n} b_i \cdot c_{n-i} \quad . \tag{13}$$

Then there is a constant a such that $a_n \le a \cdot r^{\alpha \cdot n}$.

We obtain our result.

Theorem 6. Let $F \subseteq X^\omega$ be finite-state and closed. Then $L_\alpha(F) < \infty$ if and only if

$$\exists c (c > 0 \wedge \forall n (n \in \mathbb{N} \rightarrow s_F(n) \le c \cdot r^{\alpha \cdot n})) \ .$$

Proof. In virtue of Eq. (6) the condition is sufficient. Now, suppose F to be finite-state, closed and $L_\alpha(F) < \infty$.

If $L_\alpha(F) = 0$ then according to Theorem 3 $\alpha > \dim F$ and the assertion follows from Theorem 1 .

Otherwise, consider the subset $E \subseteq F$ defined in Eq. (11). Since $F/w = F/v$ implies $E/w = E/v$, the set E is also finite-state, and since F is closed, E is also closed. Moreover, $\alpha = \dim F = \dim E$, and $L_\alpha(E/w) = r^{\alpha \cdot |w|} \cdot L_\alpha(E \cap w \cdot X^\omega) > 0$ for all $E/w \neq \emptyset$. Hence $\kappa := \min \{ L_\alpha(E/w): E/w \neq \emptyset \} > 0$, and we obtain from the fact that E is the disjoint union of the sets $E \cap w \cdot X^\omega$ $(E/w \neq \emptyset)$ the inequality

$$\infty > L_\alpha(E) \ge s_E(n) \cdot r^{\alpha \cdot n} \cdot \kappa \quad ,$$

whence $s_E(n) \le b \cdot r^{\alpha \cdot n}$ for some constant $b > 0$.

In order to link s_E to s_F , we consider those F/w for which $E/w = \emptyset$, i.e. $L_\alpha(F/w) = 0$. Then according to Theorem 3 we have $\dim F/w < \alpha$, and since F is finite-state, we get a $\gamma < \alpha$ such that $\max \{ \dim F/w: E/w = \emptyset \} < \gamma$.

Let $s: \mathbb{N} \rightarrow \mathbb{N}$ be the sum of all structure functions $s_{F/w}$ where $E/w = \emptyset$. Theorem 1 shows that there is a constant $c > 0$ such that for all $n \in \mathbb{N}$ we have $s(n) \le c \cdot r^{\gamma \cdot n}$.

Now the defining equation of E yields the inequality

$$s_F(n) \le \sum_{i=0}^{n} s_E(i) \cdot s(n-i)$$

which together with Property 5 proves the necessity of the condition. Q.E.D.

Our theorem can be applied to decide whether the (dim F)-dimensional measure of a finite-state subset $F \subseteq X^\omega$ is finite or infinite, because the condition derived there can be effectively read of from the adjacency matrix A_F of the set F .

CONCLUDING REMARK

In this paper we reduced the problem of the calculation of the Hausdorff dimension of finite-state closed subsets of X^ω to the estimation of the maximum eigenvalue of a non-negative matrix A having integer entries. Starting from a quadtree of a picture P we can, following the approach described above, construct a finite-state closed subset of Q^ω representing P and together with this set the adjacency matrix A_P from which we can estimate the Hausdorff dimension α_P of the picture P and decide whether the corresponding α_P-dimensional measure is finite or infinite.

In practice we are given a finite $2^n \times 2^n$ quadtree and before proceeding to the estimation of the Hausdorff dimension we have to extend the finite quadtree to an infinite one. This extension procedure, however, might introduce some inaccuracies into this process. These inaccuracies depend certainly on the similarity notion of quadrants chosen in the extension process, which notion is necessarily inaccurate, because we have to decide on similarity of picture quadrants on the basis a raster-picture containing finitely many pixels. The accuracy of our estimation of the Hausdorff dimension α_P of a picture P will, thus depend on the fine structure of the picture and on the number of levels of the quadtree belonging to that picture P.

The advantage of our procedure of estimating the Hausdorff dimension of a picture is that, once we have its quadtree and a suitable similarity notion of quadrants, we need not return to the picture and cover it with smaller and smaller balls, but we can instead work with the adjacency matrix A_P .

REFERENCES

[1] *Billingsley, P.: Ergodic Theory and Information. Wiley & Sons, New York 1965.*

[2] *Hunter, G.B., and K. Steiglitz: Operations on images using quadtrees. IEEE Trans. PAMI-1 (1979), 145 - 153.*

[3] *Klinger, A., and M.L. Rhodes: Organization and access of image data by areas. IEEE Trans. PAMI-1 (1979), 50 - 60.*

[4] *Lindner, R. und L. Staiger: Algebraische Codierungstheorie - Theorie der sequentiellen Codierungen. Akademie-Verlag, Berlin 1977.*

[5] *Mandelbrot, B.: Fractals, Form, Chance, and Dimnesion. Freeman, San Francisco 1977.*

[6] *Staiger, L.: Combinatorial properties of the Hausdorff dimension. In: Proc. GEOBILD '85, Wiss. Beitr. Friedrich-Schiller-Univ., Jena 1985, pp. 43 - 48.*

[7] *Staiger, L.: The entropy of finite-state ω-languages. Probl. Control Inform. Theory 14 (1985) 5, 383 - 392.*

[8] *Staiger, L.: Combinatorial properties of the Hausdorff dimension. J. Stat. Planning Inference, to appear.*

[9] *Trakhtenbrot, B.A., and Ya.M. Barzdin: Finite Automata: Behavior and Synthesis. North Holland, Amsterdam 1973.*

ON THE DISTANCE OF TWO CIRCLES

Wolfgang Ihle and Ludwig Stammler [1)]

1. Preliminaries

Let $\mathcal{K}_1, \mathcal{K}_2$ two point-sets in the n-dimensional Euklidean space, each of them being either a circle line or a circle disk. There exists uniquely their distance, the minimum of all lengths $|XY|$ $(X \in \mathcal{K}_1,\ Y \in \mathcal{K}_2)$. There are several motivations to obtain such distances, for theoretical as well as for technical aims, see [1] for instance. But it is not a simple question to find such points X,Y, neither by an usual geometrical construction nor by an elementary algebraical method, or to give complete conditions for the exact number of such pairs (X,Y). Therefore we shall only deal with some investigations and partial results in this paper.

At first let us confine ourselves to an "essential case". That means for instance, we can omit the cases $n \neq 3$: The case $n = 2$ is well known (if $\mathcal{K}_1 \cap \mathcal{K}_2 = \emptyset$, the points X,Y belong to the straight line M_1M_2, where M_i is the midpoint of $\mathcal{K}_i$), and if the planes $\alpha_i \supset \mathcal{K}_i$ span a space of dimension 4 or 5, the points $U \in \alpha_1$, $V \in \alpha_2$, for which $|UV|$ is minimal, lead without difficulty to the X,Y required. Further let us exclude the case of circle disks. This easily may be done, because if the points X,Y for two circle lines k_1, k_2 are known, it is not difficult to find the analogous points for one or two disks $\mathcal{K}_i$ instead of their boundaries k_i. Finally let us assume the circles k_1, k_2 having the same radius r. It may be indeed to obtain other results in the case of different values r_1, r_2, but the treatment of our subject and the essential methods to find those results require similar considerations.

2. A necessary condition

For each point or point set $\mathcal{P}$ in the plane α_1 let $\mathcal{P}'$ the image by orthogonal projection onto α_2, and conversely for $\mathcal{P}$ in α_2 let $\mathcal{P}'$ the image by projection onto α_1. We note at first: To a point P given in α_2 a point $X \in k_1$ has minimal distance $|PX|$ if and only if the length $|P'X|$ is minimal. (This follows immediately by $|PX|^2 = |PP'|^2 + |P'X|^2$.) Further the minimality of $|PX'|$ implies X, P', and M_1 to be collinear. (This follows in known manner by the triangle inequality.) Combining this with the same facts by changing the circles k_1, k_2, we obtain:

---

1) Dr. W. Ihle, Doz.Dr.sc. L. Stammler
Martin-Luther-Universität Halle-Wittenberg, Sektion Mathematik
DDR 4010 Halle

Theorem 1: Necessary for minimal length $|XY|$ ($X \in k_1$, $Y \in k_2$) is the property of the figure $XYX'Y'M_1M_2$ to have collinear points X, Y', M_1 as well as collinear points X', Y, M_2.

Another version of this condition may be given, if we for each $i = 1,2$ consider the "axis" s_i of k_i (the line passing perpendiculary to the plane α_i through the midpoint M_i):

Corollary: Necessary for minimal length $|XY|$ ($X \in k_1$, $Y \in k_2$) is also the property of the straight line XY, to have a nonempty intersection K with s_1 as well as a nonempty intersection L with s_2. (In special cases K or L may be the improper point of s_1 or s_2, respectively.)

Proof: In the case $YY' = s_1$ immediately holds $XY \cap s_1 \neq \emptyset$. In the case $YY' \neq s_1$ from $YY' \parallel s_1$ follows: s_1 lies in the plane passing through Y, Y', M_1. In this plane also XY lies, therefore is $XY \cap s_1 \neq \emptyset$. Analogously follows $XY \cap s_2 \neq \emptyset$.

Remark 1: It is possible to formulate this condition in an algebraical manner. But it seems becoming too complicated for obtaining essential results on an elementary geometrical or algebraical way.

Remark 2: The condition established in theorem 1 (and therefore also in the corollary) is not sufficient. For a simple counterexample let $\alpha_1 \parallel \alpha_2$, $\alpha_1 \neq \alpha_2$, and $k_1 \cap k_2' = \{X_1, X_2\}$ ($X_1 = Y_1' \neq X_2 = Y_2'$). If β is the plane passing through M_1, M_1', M_2, M_2', and $\beta \cap k_1 = \{P_1, P_2\}$, $\beta \cap k_2 = \{Q_1, Q_2\}$, then each P_iQ_j fulfils the condition, but no $|P_iQ_j|$ is minimal, rather $|X_1Y_1| = |X_2Y_2|$ is minimal (what due to the fact that $|X_1Y_1'| = |X_2Y_2'| = 0 < |P_iQ_j'|$ for all $i, j = 1,2$).

3. A first treatment of possible positions

Regarding the axes s_i mentioned in the corollary gives us occasion to discriminate the possible positions of two circles, using these axes. In this discussion we also obtain partial results of our problem.

Case 1: $s_1 \parallel s_2$. In this case is $\alpha_1 \parallel \alpha_2$, and therefore k_1', k_2' are circles too. It follows that the well-known solution of the problem concerning k_1 and k_2' leads to the solution concerning k_1 and k_2: If $k_1 = k_2'$, all $|XX'|$ ($X \in k_1$) are minimal. If $k_1 \cap k_2' = \{X_1, X_2\}$ ($X_i = Y_i'$, $X_1 \neq X_2$), the two $|X_iY_i|$ and only these are minimal, see also the above remark 2. If $k_1 \cap k_2' = \{P\}$ or $k_1 \cap k_2' = \emptyset$, the points $P \in k_1 \cap M_1M_2'$, $Q' \in k_2' \cap M_1M_2'$, lying between M_1 and M_2', fix the only minimal $|PQ|$.

Case 2: s_1 and s_2 have only one common (proper) point S. In this case the corollary above gives the solution of our problem: A straight line having nonempty intersection (proper or improper points) with s_1 and s_2, must be either passing through the point S or lying in the plane σ passing through s_1, s_2. This proves the following procedure: Let γ_1 the conus consisting of all straight lines SX ($X \in k_1$). (If $M_1 = S$, the conus becomes a plane.) The two coni γ_1, γ_2 may have two or one or no common lines. If they have two or one, on each of these there exist uniquely two points $G_n \in k_1$ and $H_n \in k_2$ ($n = 1,2$ or $n = 1$). Further let $\sigma \cap k_1 = \{E_1, E_2\}$, $\sigma \cap k_2 = \{F_1, F_2\}$. Then it is possible to select from the finitely many $|E_iF_j|$ and (if existing) $|G_nH_n|$ the minimal length, it may be of course the length of one or more amongst the pairs (E_i, F_j), (G_n, H_n). The so found pair(s) is (resp. are) the only solution(s).

Case 3: s_1 and s_2 are not lying in a common plane. This configuration unfortunately must be regarded to be the "main" case, for it holds the remark 1 above. Therefore one searches for another way in order to get further results.

4. An iterative approximation

Theorem 1 leads to the following iterative processing:

> Let $X_0 \in k_1$ an arbitrary start point. For each point $X_n \in k_1$ found before, define as follows: Let $Y_n \in k_2$ the point with minimal $|X_n'Y_n|$ (lying on the line $X_n'M_2$), and let $X_{n+1} \in k_1$ the point with minimal $|X_{n+1}Y_n'|$ (lying on the line $Y_n'M_1$).
>
> From this definition all cases with $X_n' = M_2$ or $Y_n' = M_1$ must to be excluded.

Further let us provisionally assume the resulting sequence of pairs (X_n, Y_n) ($n = 0,1,2,\ldots$) to be convergent. Obviously under this assumption the limit pair (X, Y) fulfils the condition in theorem 1. The sequence of lengths $|X_nY_n|$ by definition is decreasing and therefore convergent. But for the sequence of pairs (X_n, Y_n) some open questions are remaining: How depends the existence, the convergence (or divergence, if possible) on the start point, and how many limit pairs may exist? Which of the limit pairs (if there are more then one) give the minimal length $|XY|$?

Let us consider, before dealing with this problems, an "inverse" question. In spite of its simplicity, it demonstrates the complexity of our subject under consideration, and it calls our attention to "linked" circles (see 6.1.):

5. An inverse detail result

Theorem 2: The only case in which exist noncollinear points $X \in k_1$ and $Y_1, Y_2 \in k_2$ with minimal $|XY_1| = |XY_2|$, is the case in which $\alpha_1 \perp \alpha_2$ holds and each of the circles k_1, k_2 passes through the midpoint of the other circle.

Proof: In the case 1 mentioned above belongs to each point $X \in k_1$ at most one $Y \in k_2$ with minimal $|XY|$. Hence this case must be excluded.

We prove the same for the case 2. In this case follows from the assumption of theorem 2: It holds $X \neq S$, at least one of the lines XY_1, XY_2 does not pass through S. This line therefore (see case 2) must lie in σ. Now from $X \in \sigma$, $X \notin s_1$ and $XY_i \cap s_1 \neq \emptyset$ follows $Y_i \in \sigma$ $(i = 1,2)$, this implies Y_1', Y_2', X, M_1 to be collinear, lying in $g = \sigma \cap \alpha_1$. For any $Z \in k_2, \notin \sigma$ follows $Z' \notin g$, the ray M_1Z' must have with k_1 an intersection $\tilde{X} \neq X$. This and $X \in s_2$ (proved by $X \in \sigma$ and $|XY_1| = |XY_2|$) implies the contradiction $|\tilde{X}Z| < |XZ| = |XY_1|$.

It remains case 3. Let UV ($U \in s_1$, $V \in s_2$) be the common normal line of s_1, s_2. It holds $X \notin s_1$. The case $X \notin s_2$ cannot occur, because there exists in this case only one line passing X and having nonempty intersection with s_1 and s_2, and therefore the condition of the corollary cannot be fulfiled by Y_1 and Y_2. Consider now the case $X \in s_2$ (that implies $|XZ| = |XY_1|$ for all $Z \in k_2$) and $X \neq V$. The plane passing through U,V,X is orthogonal to s_1 and therefore equal to α_1. This implies $M_1 = U$ and $s_2 \subset \alpha_1$ and $r = |M_1X| > |M_1V|$. From this and $M_2 \in s_2$ follows: k_2 must have a point Z lying on a line l in distance r parallel to s_2, such that M_1 lies between s_2 and l. This is a contradiction, because to each point $Z \in l \cap k_2$ exists a point $\tilde{X} \in k_1$ with $|XY_1| = |XZ| > |\tilde{X}Z|$ (Z lies not on the ray M_1X).
Now we have $X = V$. From $X \in \alpha_1 \perp s_1$ also follows $M_1 = U$ and $s_2 \subset \alpha_1$, but now $r = |M_1V|$, and the only possibility to avoid the above contradiction (with $Z \in k_2$ on the line $l \parallel s_2$ through M_1) is $Z = M_1$, q.e.d.

6. Two "extremal" detail results

The following detail results concern to two special (but typical) "extremal" positions of the circles: At first two circles "linked" with another, at second two circles "wide removed" from another.

6.1. Let $M_2 \neq M_1$ an interior point of the circle disk bounded by k_1 (say $|M_1M_2| = e < r$), and let $\alpha_1 \neq \alpha_2$, hence $\alpha_1 \cap \alpha_2$ be the line M_1M_2.

We think the angle $\varphi = \sphericalangle(\alpha_1, \alpha_2)$ running through the interval $0 < \varphi \leq \frac{\pi}{2}$. An usual argument applying continuity shows: For small φ an optimal pair (X_1,Y_1) must lie in the neighbourhood of a point in $k_1 \cap \tilde{k}_2$, where $\tilde{k}_2$ is the limit circle, $k_2 \to \tilde{k}_2$ for $\varphi \to 0$. But for increasing φ the optimal pair must "jump" into a pair (X_0,Y_0) with X_0 , $Y_0 \in M_1M_2$. A more precise description of this "jumping" is to exist infinite many optimal pairs (X,Y) as follows:

> Theorem 3: If under the described assumptions a pair (X_1,Y_1) fulfils the condition of theorem 1 and has the length $|X_1Y_1| = |X_0Y_0|$, then each $X \in k_1$ is the first point in a pair (X,Y) with the same properties.

Proof: Let k_1 the circle $x^2 + y^2 = r^2$, $z = 0$, further $M_2 = (e,0,0)$ and α_2 spanned by $\boldsymbol{i} = (1,0,0)$, $\boldsymbol{j} = (0,f,g)$ ($f = \cos\varphi$, $g = \sin\varphi$). For each $X = (c\cdot r, s\cdot r, 0)$ $(c^2+s^2 = 1)$ we have $X' = c\cdot r\cdot\boldsymbol{i} + f\cdot s\cdot r\cdot\boldsymbol{j}$. The ray M_2X' has with k_2 the intersection $Y = (e + \frac{r}{w}\cdot(c\cdot r - e))\cdot\boldsymbol{i} + \frac{r}{w}\cdot f\cdot s\cdot r\cdot\boldsymbol{j}$ $(w = \sqrt{(cr-e)^2 + f^2s^2r^2}$), and so becomes $Y' = (e + \frac{r}{w}\cdot(cr-e), \frac{f^2sr^2}{w}, 0)$. Now an easy calculation shows: By the choice $g = \frac{e}{r}$, for each X the points M_1,X,Y' are collinear (that fulfils the condition of theorem 1) and the equation $|XY| = e$ holds true.

6.2. For the iteration described above (see 4.) and their convergence we have:

> Theorem 4: For the convergence of the sequence (X_n) into an unique limit point X, not depending on the start point $X_0 \in k_1$, the following assumption of a "wide separated" position is sufficient: The ellipse k_1' totally lies outside of k_2, and k_2' totally lies outside of k_1.
> Especially follows: In this position the circles have an unique pair (X,Y) with minimal $|XY|$.

Remark: Many practical numerical experiments show: The iteration is also convergent in other cases (often with a good speed), but the dependence on the start point may be more complicated.

Proof of theorem 4: For n = 0,1,2,... we consider the arc C_nD_n of possible positions for Y_n, when the start point X_0 runs along the whole circle line k_1, and the (straight) length $\eta_n = |C_nD_n|$ belonging to this arc. Analogously let $\eta_n' = |C_n'D_n'|$, $\xi_{n+1} = |A_{n+1}B_{n+1}|$, $\xi_{n+1}' = |A_{n+1}'B_{n+1}'|$ for Y_n', X_{n+1}, X_{n+1}', respectively. For n = 1,2,3,... from the relation between a circle and his affin image follows $\xi_n' \leq \xi_n$. Let p_1 the minimal distance $|M_2Z|$ $(Z \in k_1')$. Then for $q_1 = \frac{r}{p_1} < 1$ (see the assumption), from $r = |M_2C_n| = |M_2D_n|$ and $p_1 \leq |M_2A_n'|, |M_2B_n'|$ follows $\eta_n \leq q_1 \cdot \xi_n'$. Analogously holds $\eta_n' \leq \eta_n$ and $\xi_{n+1} \leq q_2 \cdot \eta_n'$ with a positive $q_2 < 1$. From $\xi_{n+1} \leq q_1 \cdot q_2 \cdot \xi_n$ follows $\xi_n \to 0$, this completes the proof.

Literature

[1] Hauschild,W., M.Dietrich, W.Schufft, and H.Schwab: Teller-Segmentelektroden für Hochspannungsprüfanlagen. ELEKTRIE 37 (1983) 4.

[2] Nagel,U.: Geometrische Berechnungen an Tellersegmentelektroden. Belegarbeit (Manuskript) VEB Transformatoren- und Röntgenwerk "Hermann Matern" Dresden (1986).

[3] Stammler,L.: Algorithmen zu toroidalen Kreisscheibensystemen. Studie (Manuskript) Martin-Luther-Universität Halle-Wittenberg, Sektion Mathematik (1985).

[4] Ueltzen,C.: Entwurf und Berechnung von Stützgerüsten für toroidförmige Segmentelektroden. Belegarbeit (Manuskript) VEB Transformatoren- und Röntgenwerk "Hermann Matern" Dresden (1985).

PRECISION OF SYSTEMATIC SAMPLING ON A STEP FUNCTION

Luis M. Cruz-Orive[1)]

SUMMARY

The basic problem considered is to predict the variance of the estimator of the integral of a step function obtained by systematic sampling. Using Matheron's theory of 'regionalized variables' a simple approximation is obtained for the trend of the required variance. The result is applied to estimating the total linear projection of a plane curve by intersection counting with a test system of parallel lines. The variance prediction coincides with that derived by Moran (1966) using a different but related route. The problem of sampling from a discrete population (related to the fractionator method of stereology) is also briefly examined.

Keywords: Covariogram; Curve length; Fractionator; Monte-Carlo integration; Step function; Stereology; Systematic sampling; Variance of estimator.

1. INTRODUCTION

The basic problem considered in this paper is to estimate the integral of a step function of bounded variation, defined on a bounded domain of the real line, by means of systematic sampling. The estimation method can be regarded as a Monte-Carlo integration method with systematic sampling.

Using the classical representation of the variance of the estimator in terms of the Fourier transform of the covariogram of the step function, an explicit approximation is derived for the trend of the estimator variance (section 2). The approach is inspired by Matheron's transitive theory or regionalized variables (Matheron, 1965, 1971, Serra, 1967; for a survey see Cruz-Orive, 1988).

A direct application of the basic result concerns the estimation of the length of a plane curve by intersection counting with a test system of parallel lines (section 3). We consider only the variance of the estimator of the total projection of the curve onto a fixed axis, namely the conditional variance of the length estimator for a given orientation of the test system. The result coincides with that obtained by Moran (1966) by a different but related route. In the latter paper the variance component due to the isotropic orientation of the test system is also considered - see subsection 3.4 here.

As a related problem we consider the estimation of the total of a parameter defined on a finite, discrete population of objects, or on a partition of a population defined on a bounded subset of $\mathbb{R}^d$, (Jolly, 1979). In the stereological context the estimation procedure is known as the 'fractionator' (Gundersen, 1986). See section 4.

---

1) Anatomisches Institut der Universität Bern, Postfach 139, CH-3000 Bern 9

2. ESTIMATION OF THE INTEGRAL OF A STEP FUNCTION

2.1. Systematic sampling along $\mathbb{R}^1$

Let f(x) be a function of bounded variation over a compact domain $D \subset \mathbb{R}^1$. We assume that f(x) is Lebesgue-integrable in D and $f(x) \equiv 0$ for $x \notin D$. The problem is to estimate the deterministic, (i.e. non-random), quantity:

$$Q = \int_D f(x)dx. \tag{2.1}$$

We sample f(x) at the points of a lattice placed uniformly at random over D. Thereby we get a systematic sample $\{f(z + kT),\ k \text{ integer}\}$, where T is the fixed distance beween test points, $(0 < T < \infty)$ and z is a uniform random variable in the interval [0, T]. The estimator

$$\text{est}Q = T\sum_k f(z + kT) \stackrel{d}{=} q(z) \tag{2.2}$$

is easily shown to be unbiased for Q. The following representation of the variance of the preceding estimator is well known (for early references see Kendall & Moran, 1963, section 5.4):

$$\text{Var}q(z) = 2\sum_{k=1}^{\infty} G(k/T) \tag{2.3}$$

where G is the Fourier transform of the covariogram g of f, namely:

$$G(t) \stackrel{d}{=} \mathcal{F}g = \int_{-\infty}^{\infty} g(h)\exp(-2\pi ith)dh, \quad (-\infty < t < \infty) \tag{2.4}$$

$$g(h) \stackrel{d}{=} f*\check{f} = \int_{-\infty}^{\infty} f(x)f(x+h)dx, \quad (-\infty < h < \infty). \tag{2.5}$$

Matheron (1965, 1971) proposes to replace g(h) with a positive definite model in the form of a limited expansion near the origin. In fact, the behaviour of g(r) near $r \stackrel{d}{=} |h| = 0$ explains mainly the trend of Varq(z), whereas any oscillations of this variance depend upon the behaviour of g(r) for larger values of r. In practice, only the estimation of the trend is accessible, in general.

A useful subclass of models for g(r) near the origin is the polynomial:

$$g(r) = b_0 + b_1 r + b_2 r^2 + \dots \tag{2.6}$$

under which the trend of Var q(z) is approximated by

$$\text{Var}q(z) = -\sum_{j=1}^{\infty} b_{2j-1} j^{-1} B_{2j} T^{2j} \tag{2.7}$$

where B_{2j} is the Bernoulli number

$$B_{2j} = (-1)^{j-1} 2(2j)!(2\pi)^{-2j} \sum_{k=1}^{\infty} k^{-2j},$$

(e.g. $B_2 = 1/6$, $B_4 = -1/30$, $B_6 = 1/42$, ...). Retaining only the linear term and noting that $b_1 = g'(0)$ we get

$$\mathrm{Var}q(z) \approx -(1/6)g'(0)T^2, \tag{2.8}$$

which is the key approximation used below. For useful applications see Serra (1967), Digabel (1971), Thioulouse et al. (1985), Gundersen & Jensen (1987), Michel & Cruz-Orive (1988).

2.2. Systematic sampling on a step function

Let f(x) be a step function of bounded variation defined as follows:

$$f(x) = \begin{cases} I_j, \text{ constant, for } x \in (x_{j-1}, x_j], & (j=1,2,\dots,N, \quad N \text{ finite}), \\ 0 \quad \text{for } x \notin D, \end{cases} \tag{2.9}$$

$$D = \bigcup_{j=1}^{N} (x_{j-1}, x_j], \qquad (0 = x_0 < x_1 < \dots < x_N < \infty).$$

The target quantitiy Q, see (2.1), can be written in this case as follows:

$$Q = \sum_{j=1}^{N} (x_j - x_{j-1}) I_j, \tag{2.10}$$

which is unbiasedly estimated by the right hand side of (2.2). To predict the variance of this estimator using (2.3) we need the covariogram of f near the origin. For r near 0 we have:

$$f(x)f(x+r) = \begin{cases} I_j^2 & \text{if } x \in (x_{j-1}, x_j - r], \\ I_j I_{j+1} & \text{if } x \in (x_j - r, x_j], \end{cases} \tag{2.11}$$

for j = 1,2, ..., N. Thus, for r near 0,

$$\begin{aligned} g(r) &= \int_{-\infty}^{\infty} f(x)f(x+r)dx \\ &= \sum_j \{\int_{x_{j-1}}^{x_j - r} I_j^2 dx + \int_{x_j - r}^{x_j} I_j I_{j+1} dx\} \\ &= \sum_j (x_j - x_{j-1}) I_j^2 - r\sum_j I_j^2 + r\sum_j I_j I_{j+1}, \end{aligned} \tag{2.12}$$

which is a polynomial of order 1. Applying (2.8) we get the following approximation for the trend of the required variance:

$$\mathrm{Var}(\mathrm{est}Q) = \frac{1}{6}\left(\sum_{j=1}^{N} I_j^2 - \sum_{j=1}^{N-1} I_j I_{j+1}\right)T^2, \tag{2.13}$$

which admits the alternative, more useful expression:

$$\mathrm{Var}(\mathrm{est}Q) = \frac{1}{12}\left(I_1^2 + I_N^2 + \sum_{j=1}^{N-1} \Delta_j^2\right)T^2, \tag{2.14}$$

where $\Delta_j = I_{j+1} - I_j$ is the height of the j-th 'jump' of the step function, $j = 1,2, \ldots, N-1$.

An immediate practical consequence of the preceding result is that the precision of the estimation of the integral Q of a step function by systematic sampling is higher the smaller the first and last values (I_1, I_N, respectively) and the jumps $\Delta_1, \ldots, \Delta_{N-1}$. Thus, optimal shapes for the graph of the step function are dome-like shapes (eventually with a plateau at the top).

3. APPLICATION: ESTIMATING THE LENGTH OF A PLANE CURVE BY INTERSECTION COUNTING.

3.1. Background

Consider a plane rectifiable curve of finite length L. An efficient way of estimating L is by intersection counting with a system of parallel lines at a distance T. The position of the test system on the plane is determined by the coordinates (z,ω) of one of the lines, where z is the distance from a fixed origin to the line $(0 < z \leqslant T)$ and ω the angle between an axis $l(\omega)$ normal to the line and a fixed axis, $(0 < \omega \leqslant \pi)$.

If z, ω are uniform random in $(0, T]$ and $(0, \pi]$, respectively, and independent, that is, if the test system is Isotropic Uniform Random (IUR) on the plane, then it is easy to show that

$$\text{est}L = (\pi/2)\cdot T\cdot I(z,\omega) \tag{3.1}$$

is an unbiased estimator of L, ('Buffon's estimator'! see e.g. Santaló, 1976), where $I(z,\omega)$ is the total number of intersection points between the curve and the test system for each pair (z,ω). The variance of estL can be expressed as follows:

$$\begin{aligned}(4/\pi^2)\text{Var}(\text{est}L) &= \text{VarE}\{TI(z,\omega)|\omega\} + \text{EVar}\{TI(z,\omega)|\omega\} \\ &= \text{Var}Q(\omega) + \text{EVar}\{\text{est}Q(\omega)|\omega\}\end{aligned} \tag{3.2}$$

where $Q(\omega)$ represents the length of the total orthogonal projection of the curve onto the axis $l(\omega)$, see Fig. 1 and Kendall & Moran (1963, section 3.8). For fixed ω and $z \sim U(0, T]$,

$$\text{est}Q(\omega) = T\cdot I(z,\omega) \tag{3.3}$$

is an unbiased estimator of $Q(\omega)$ of the form (2.2).

Moran (1966) considers in turn the two variance components in the right hand side of (3.2). Here we concentrate on the conditional variance of $\text{est}Q(\omega)$ only. First we outline Moran's derivation (subsection 3.2) and then (subsection 3.3) we show that his result follows directly from formula (2.14). In subsection 3.4 we give a practical formula, involving both variance components, for Var(estL) when a square grid test system is used.

3.2. Moran's approach

Writing $Q(\omega) \equiv Q$, for short, the unbiased estimator (3.3) of Q can be expressed as (2.2). The function f(x) represents the number of intersections between the curve and a test line normal to the projection axis $l(\omega)$ at a point of abscissa x measured on that axis, (see Fig. 1).

Instead of (2.3), Moran uses the equivalent expression:

$$\mathrm{Var}(\mathrm{est}Q) = 2 \sum_{k=1}^{\infty} |F(k/T)|^2 \tag{3.4}$$

where F is the Fourier transform of f namely

$$F(t) \stackrel{d}{=} \mathcal{F}f = \int_{-\infty}^{\infty} f(x)\exp(-2\pi itx)dx, \qquad (-\infty < t < \infty),$$

(note that $G(t) = \mathcal{F}f*\check{f} = (\mathcal{F}f)(\overline{\mathcal{F}}f) = |\mathcal{F}f|^2$).

Let $x(s)$, $(0 \leqslant s \leqslant L)$, denote the abscissa of a point which describes the curve, and set $x(0) \equiv x_0 = 0$, $x(L) \equiv x_N$, where N is a finite positive integer. If the curve is closed, then $x_N \equiv x_0$. The successive abscissae at which $x(s)$ changes direction (i.e. the turning abscissae) are denoted by $x_1, x_2, \ldots, x_{N-1}$, see Fig. 1. Clearly

$$F(t) = \int_{x_0}^{x_1} \exp(-2\pi itx)dx + \int_{x_2}^{x_1} \exp(-2\pi itx)dx + \ldots$$

$$= (2\pi it)^{-1}\{\exp(-2\pi itx_0) - 2\exp(-2\pi itx_1) + 2\exp(-2\pi itx_2) - \ldots\}, \tag{3.5}$$

the last term in the preceding summation being $\exp(-2\pi itx_0)$ if the curve is closed and $\pm \exp(-2\pi itx_N)$ if $x_N \neq x_0$. To make use of (3.4), Moran notes that each term in the expansion of $|F(k/T)|^2$ is periodic in the x_j with period T, and thereby he regards each x_j as a uniform random variable in the interval $(0, T]$. In the corresponding expectation of $|F(k/T)|^2$ the cross-products cancel out because $E\{\exp(-2\pi ikx/T)\} = 0$ for $k = 1,2,\ldots$. If the curve is 'open' (i.e. $x_N \neq x_0$) then, for $t = k/T$,

$$E|F(t)|^2 = (4\pi^2t^2)^{-1}\{E|\exp(-2\pi itx_0)|^2 + 4E|\exp(-2\pi itx_1)|^2 + 4E|\exp(-2\pi itx_2)|^2 + \ldots\}$$

$$= (4\pi^2t^2)^{-1}\{1 + 4(N-1) + 1\}. \tag{3.6}$$

If the curve is closed, then

$$E|F(t)|^2 = (4\pi^2t^2)^{-1} \cdot 4N. \tag{3.7}$$

Replacing $|F(k/T)|^2$ in the right hand side of (3.4) with the preceding expectations, Moran obtains, in turn:

$$\mathrm{Var}\{\mathrm{est}Q(\omega)|\omega\} = \begin{cases} (1/6)(2N-1)T^2 & \text{if the curve is open,} \quad (3.8) \\ (1/3)NT^2 & \text{if the curve is closed.} \quad (3.9) \end{cases}$$

Remark. Consider a sweeping line perpendicular to the projection axis (i.e. parallel to the test lines) and let n(tan) denote the number of tangential positions of the sweeping line (possibly at a cusp) with the curve, but excluding its end-points. Then the integer N in (3.8), (3.9) equals:

$$N = n(\tan) + 1 \quad \text{if the curve is open,} \tag{3.10}$$

$$N = n(\tan) \quad \text{if the curve is closed.} \tag{3.11}$$

(see Fig. 2).

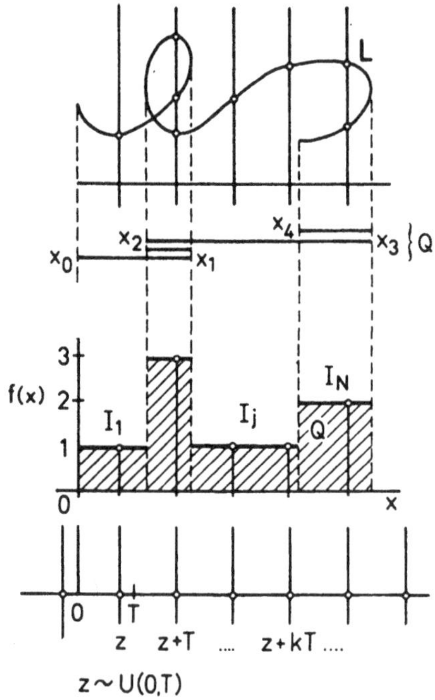

Fig. 1

Fig. 1. The step function f(x) = "number of intersections between a bounded curve and a test line perpendicular to a given axis at a point of abscissa x." The total projection Q of the curve onto the axis equals the integral of f(x), see section 3.

Fig. 2. Basic types of the step function of intersections. (a,b,c): curve with two end-points. (d): closed curve. See subsection 3.3.

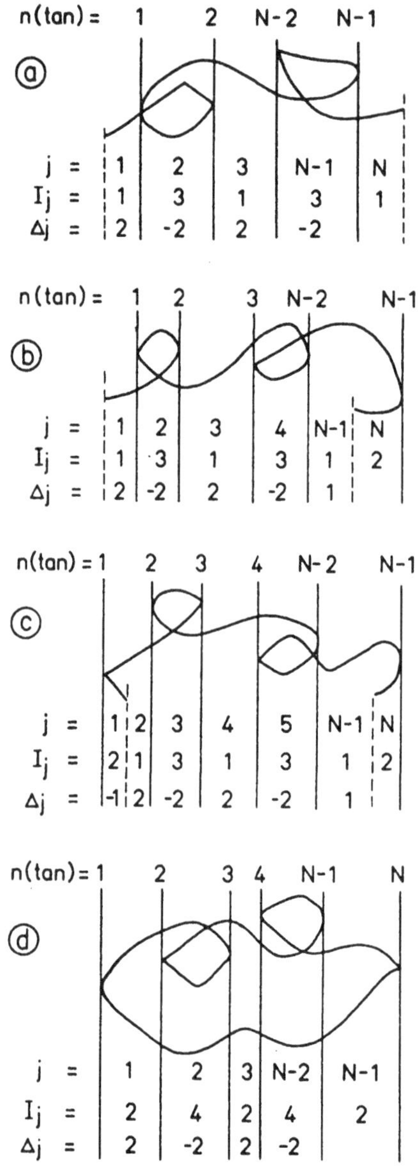

Fig. 2

3.3. The step function approach

If the turning abscissae mentioned above are re-labelled so that $0 = x_0 < x_1 < x_2 < \ldots < x_N$, then $f(x)$ is a step function of the form (2.9), where I_j is a bounded non-negative integer representing the number of intersections between the curve and a test line normal to the projection axis at a point whose abscissa lies in the interval $(x_{j-1}, x_j]$, $(j = 1,2,\ldots,N)$, see Fig. 1. We want to verify that the right hand side of (2.14) coincides in this case with either (3.8) or (3.9).

The curve is open (i.e. $x_N \neq x_0$). The different cases arising here can be essentially reduced to the three cases illustrated in Fig. 2. If $n(\tan) = N - 1$, see (3.10), then the number of intervals of the step function $f(x)$ is almost surely N.

Case (a): $I_1 = I_N = 1$ and $N - 1$ jumps Δ_j are equal to ± 2.

Case (b): $I_1 = 1$, $I_N = 2$, (or $I_1 = 2$, $I_N = 1$); $N - 2$ jumps are equal to ± 2 and one jump is equal to ± 1.

Case (c): $I_1 = I_N = 2$; $N - 3$ jumps are equal to ± 2 and two jumps are equal to ± 1.

In all three cases, the right hand side of (2.14) reduces to (3.8), as we wanted to verify.

The curve is closed. If $n(\tan) = N$, see (3.11), then the number of intervals of $f(x)$ is almost surely $N - 1$, see Fig. 2(d). In this case $I_1 = I_{N-1} = 2$ whereas the jumps Δ_j are equal to ± 2, $(j = 1, 2, \ldots, N - 2)$. Therefore the right hand side of (2.14) coincides with (3.9), as expected.

3.4. Practical estimation of the total length L of a finite collection of plane curves using a square grid of side T.

By extension of equ. (3.1), an unbiased estimator of L for a square grid is:

$$\text{est}L = \frac{\pi}{4} \cdot T \cdot I \tag{3.12}$$

where I is the total number of intersections between the curves and the lines of the grid. For a single curve, the first variance component in the right hand side of (3.2) if given by Moran (1966, Table 2) - see below. Assuming independence between second-component errors along each direction of the grid we get:

$$CE^2(\text{est}L) \approx 0.003869 + \{n_1(\tan) + n_2(\tan) + c\}/(3I^2), \tag{3.13}$$

where $n_1(\tan)$, $n_2(\tan)$ denote the respective numbers of 'tangents', (excluding those at curve end points) parallel to the two directions of the test lines, whereas $c = 0$ if the curve is closed and $c = 1$ if the curve is open.

For a finite collection of m rectifiable curves, the first term in the right hand side of equ. (3.13) remains unchanged, whereas the quantities $n_1(\tan)$, $n_2(\tan)$, $\underline{c}$ and $\underline{I}$ have to be replaced with the corresponding sums over the m curves.

4. SYSTEMATIC SAMPLING FROM A DISCRETE POPULATION OF OBJECTS

Consider a finite collection of bounded real numbers $\{W_1, W_2, \ldots, W_N\}$. The problem is to estimate the total

$$Q = W_1 + W_2 + \ldots W_N. \tag{4.1}$$

For a fixed positive integer T, an unbiased estimator of Q is:

$$\text{est}Q = T \cdot \sum_k W_{[z]+kT} \tag{4.2}$$

where k runs over the integers, [z] is a uniform random integer from the set $\{1, 2, \ldots, T\}$ and $W_j \equiv 0$ for $j \notin \{1, 2, \ldots, N\}$.

To predict Var(estQ) it is tempting to apply the results in subsection 2.2 with $x_j - x_{j-1} = 1$ for all $j = 1,2,\ldots,N$ and $I_j \equiv W_j$. This was done in Cruz-Orive (1987, page 157), with useful results for practical purposes. From a more strict viewpoint, however, the result (2.14) is not readily applicable here. For instance, $f(x + r)$ is not defined for $0 \leqslant r < 1$ and hence the construction (2.11) does not make sense.

Instead of starting from the relation (2.3), we may start from the alternative, also exact representation:

$$\text{Var}(\text{est}Q) = Tg(0) + 2T \sum_{k=1}^{\infty} g(kT) - Q^2 \tag{4.3}$$

where

$$g(r) = \sum_{j=1}^{N-r} W_j W_{j+r}, \qquad r = 0, 1, 2, \ldots \tag{4.4}$$

Thus, we get

$$\text{Var}(\text{est}Q) = T \sum_{j=1}^{N} W_j^2 + 2T \sum_{k=1}^{\left[\frac{N-1}{T}\right]} \sum_{j=1}^{N-kT} W_j W_{j+kT} - Q^2, \tag{4.5}$$

which allows exact variance computations for a given collection of numbers $\{W_1, W_2, \ldots, W_N\}$.

A simple approximation of the preceding expression (in the spirit of (2.14), but eventually improving it) would be very useful.

ACKNOWLEDGEMENTS

I wish to thank Rose-Marie Fankhauser for typing the manuscript and Karl Babl for preparing the figures. (Work supported by the Swiss National Science Foundation grant no. 3.487-0.86).

REFERENCES

Cruz-Orive, L.M. (1987) Precision of stereological estimators from systematic probes. Proceedings of 7th International Congress for Stereology, F-Caen 1987 (ed. by J.L. Chermant). Acta Stereol. 6/III, 153-158.

Cruz-Orive, L.M. (1988) On the precision of systematic sampling: a review of Matheron's transitive methods. Proceedings of the Fourth International Conference in Stereology and Stochastic Geometry, Berne 1987. J. Microsc., in press.

Digabel, H. (1971) Estimation du volume ovogonial d'une gonade. Internal Report no. 259, Ecole Nationale Supérieure des Mines de Paris, F-Fontainebleau.

Gundersen, H.J.G. Stereology of arbitrary particles. A review of unbiased number and size estimators and the presentation of some new ones, in memory of William R. Thompson. J. Microsc. 143, 3-45.

Gundersen, H.J.G. and Jensen, E.B. (1987) The efficiency of systematic sampling in stereology and its prediction. J. Microsc. 147, 229-263.

Jolly, G.M. (1979) Sampling of large objects. In: Sampling Biological Populations. Cormack, R.M., Patil, G.P. and Robson, D.S. (eds.), Fairland, Maryland: International Co-operative Publishing House, 193-201.

Kendall, M.G. and Moran, P.A.P. (1963) Geometrical Probability. Griffin, London.

Matheron, G. (1965) Les Variables Régionalisées et Leur Estimation (Thesis). Masson et Cie, Paris.

Matheron, G. (1971) The Theory of Regionalized Variables and Its Applications. Les Cahiers du Centre de Morphologie Mathématique de Fontainebleau, No. 5. Ecole Nationale Supérieure des Mines de Paris. F-Fontainebleau.

Moran, P.A.P. (1966) Measuring the length of a curve. Biometrika 53, 359-364.

Michel, R.P. and Cruz-Orive, L.M. (1988) Application of the Cavalieri principle and vertical sections method to lung: estimation of volume and pleural surface area. J. Microsc. 150, 117-136.

Santaló, L.A. (1976) Integral Geometry and Geometrical Probability. Addison-Wesley, Reading, Massachusetts.

Serra, J. (1967) Echantillonnage et Estimation Locale des Phénomènes de Transition Miniers. Thèse de Docteur-Ingénieur. Institut de Recherches de la Sidérurgie (IRSID). Station d'Essais, F-Maizières-lès-Metz.

Thioulouse, J., Mathy, B. and Ploye, H. (1985) Estimation d'un volume a partir de coupes seriées: sous-echantillonnage, covariogramme transitif et calcul de précision. Mikroskopie 42, 215-224.

Applications of Computational Geometry in Knowledge-Based Dialog Systems

Doron Nussbaum[1] Jörg-Rüdiger Sack[1] Thomas Strothotte[2]

Abstract

Knowledge-based dialog systems which use diagrams as well as natural language in explanations to users require advanced techniques of computational geometry to convert between pictorial and fregean forms of knowledge representation. This, in turn, gives rise to new geometric problems which must be solved. In this paper we illustrate the processing necessary in such dialog systems by motivating and solving a specific geometric problem.

1 Introduction

The purpose of knowledge-based dialog systems is to allow the user to solve problems in a limited domain of discourse interactively with a machine. The user specifies his problem in such a way that the system can arrive at the solution by carrying out inferences based on its knowledge of the domain. This knowledge is typically specified in one of various symbolic (or *fregean*) forms, each of which is essentially equivalent to Gottlob Frege's predicate calculus [1]. The programming language *Prolog*, for example, allows a simple implementation of such knowledge representation. This kind of representation of knowledge results in the human-computer dialogs being primarily *language-oriented*. The dialog is often carried out in a natural language or one which resembles a natural language. Furthermore, the concepts which are used in the dialogs tend to be those for which terms in a natural language exist.

The goal of our work in dialog systems has been to integrate pictures and natural language in human-computer dialogs. Our experience has shown that this is only possible if the knowledge base contains not only fregean symbols but also pictorial representations. Thus certain pieces of information are represented only symbolically while others are represented only pictorially. (Some may be doubly represented, but these will not be of concern here.)

The presence of such forms of representation leads to the requirement that a conversions from one form to another be possible. As an example, suppose the user of a dialog system concerned with the use and maintenance of PC products ask the question, "Where is the intensity adjusted?" To answer this question, an appropriately set up system can present the user with the diagram shown in Figure 1, where the intensity switch on the monitor is highlighted. This picture may be a very good answer for some users, but a verbal form of the answer may be necessary in some situations, either to be passed directly on to the user or because further computations are to be carried out on this answer. One such verbal form may be "On the second switch from the left on the bottom of the monitor." For the machine to arrive at such a formulation starting with the diagrammatic representation of the answer, however, requires complex geometric processing of diagrams.

Figure 1: Sample diagram

[1]School of Computer Science, Carleton University, Colonel By Drive, Ottawa, Ontario K1S 5B6, Canada
[2]Abteilung Dialogsysteme, Institut für Informatik, Universität Stuttgart, Azenbergstr. 12, D-7000 Stuttgart 1, F. R. Germany

2 A Software-Architecture

We have designed a new software-architecture for knowledge-based systems in which pictorial and fregean representations of knowledge are integrated (see Figure 2). The application program communicates with a reasoning component which draws on pictorial and fregean knowledge. A *picture interpretation component* (PIC) handles the conversions between these two forms. The component can be used bi-directionally: if the input is a symbolic representation then the output is a pictorial one, while if the input is a pictorial one, the output is symbolic.

The information flowing counterclockwise from the reasoning component to the PIC corresponds to generating pictures from fregean representations, while information flowing clockwise corresponds to extracting from pictures information in a fregean form. The former transformation is described elsewhere (see [2,9]); the latter transformation – from pictorial to symbolic forms – requires that the PIC have algorithms which analyse a pictorial representation to determine the correctness of certain conditions. This requires sophistocated analysis using techniques from computational geometry. In the next section of this paper, we illustrate such analysis with one specific problem. The problem is to find so-called *chains*, which are regular arrangements of objects in a diagram; a formal definition follows below. The presence of such chains must be determined so that a knowledge-based system can correctly produce such descriptions as "the leftmost ..." or "the second ... from the right", etc.

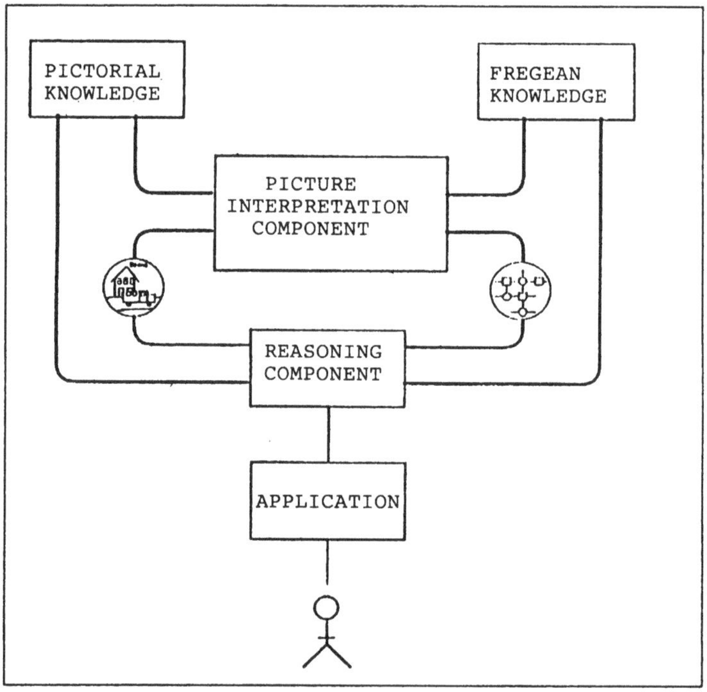

Figure 2: Software-Architecture

3 Algorithms

3.1 Relationships between pairs of objects

We assume that the diagram is represented as a set of polygons. We proceed by introducing some notation. Let P be a simple polygon. We denote by bd(P) the boundary of P; furthermore, int(P) is defined as P – bd(P). Let d be a direction represented as an oriented line. Then proj(A) is the perpendicular projection of A onto d. Let A, B be two objects represented as simple polygons. Then we say that

- A and B are comparable in direction d if proj(A) intersects proj(B).
- A and B are linearly separable in a direction d if there exists a line g parallel to d such that A and B are located in opposite half-planes defined by g.
- A and B are comparably separable in d if A and B are comparable in d and A and B are separable in d.
- A is above B if and only if for every vertical cut through int(A) all points of A that are on the cut have larger y-value than the points of B if any.
- A is completely above B if and only if A is above B and every vertical cut intersecting A also intersects B.

In an analogous fashion we can define A is to the left of B, to the right of B etc.

3.2 Testing these relationships

It is obvious how to test two objects for comparability in a given direction. The discussion is thus omitted here. To test two objects for linear separability one need only test whether their convex hulls, CH, are non-intersecting. This is true since two polygons A and B are linearly separable if and only if CH(A) and CH(B) are non-intersecting.

If we want to determine all lines of linear separability for A and B we do the following. First we find the convex hulls of A and B, which can be done in linear time. Then we determine the two separating supporting lines for A and B. A line is a *separating supporting line* g for A and B if g intersects A and B such that A and B lie on opposite half-planes defined by g. The two separating supporting lines define a wedge which represents all lines of linear separability between A and B. This step requires logarithmic time. Thus the total time is linear in the number of vertices of A and B.

The slightly more interesting property to test for is the *above* relation. To test this relation, we define that A and B are *separable* in the y direction if B can be translated by an arbitrary amount in the direction parallel to the negative y-axis without colliding with A, or vice versa. Now observe that A is above B if and only if A and B are separable in the y-direction (not necessarily linearly separable). Thus the problem of testing the above-relation between two polygons is expressed as a translation separability problem for which a linear time solution is available [8]. It is easily seen that the approach can be slightly modified to test the *completely above* relation.

Lemma 3.1 *The* above *relation can be tested in linear time for any pair of simple polygons*

The relationships can also be generalized, in the natural way, to three dimensions. To test the relations, results on three dimensional separability are useful. For this, the interested reader is referred to the recent papers [4,5].

3.3 Finding Chains in Diagrams

We define an ordered chain $C = (P_1, P_2, ..., P_M)$ as an M-tuple of n-vertex polygons such that

(i) each pair of polygons in C is comparable in the same direction d

(ii) the entire set $S = \{P_1, P_2, ..., P_M\}$ is separable in the direction perpendicular to d.

A subset S' of S is a chain in S if S' is a chain and no polygon in S' is comparable in d with any polygon in $S - S'$.

Assume for this discussion that d is the y-direction. To test a given set S' of M' polygons for comparability in y, can be done in $O(Mn)$ time simply as follows. First find the endpoints of each polygon (endpoint = points with extremal coordinates in y-direction). Each polygon is conveniently represented by the line-segment joining its endpoints. Next project the line segments onto the y-axis. This defines a set of intervals, possibly overlaping, located on the y-axis. The total time required so far in $O(Mn)$. Now we carry out a scan of the endpoints in decreasing order of their y-values, placing a polygon into a data structure T when its upper endpoint is encountered and removing it when the lower endpoint is encountered. For S' to be a chain in S there must exist a value, v, such that all polygons in S' are intersected by a horizontal line whose y-value equals v; we call this test step (a). Furthermore, one must ensure that when the scan reaches the first endpoint of S' the tree is empty (step (b)), and no insertion of a polygon in $S - S'$ occurs before the last element of S' is deleted from T (step (c)). Executing steps (b) and (c) requires $O(M)$ time; step (a) can be performed in $O(M')$ time.

Our next task is to find all maximal subsets S' of S such each pair of polygons in S' is comparable in d. An simple $O(M \log M)$ algorithm can be designed for solving this problem based on the plane-sweep paradigm (see e.g. [6] or [7]; we assume the reader to be familiar with the details of this technique). Notice that no polygon can be part of two lists for which at least one polygon is different. Thus the total size of all sets S' is linear in the cardinality, M, of S. (Since interval overlap reduces to this problem the algorithm is also optimal.)

For the remainder, we assume that S is a set for which we know that all pairs of polygons in S are comparable. Next we address the hardest step of the chain-finding procedure which is to test whether a given subset S' of S is separable. We do this in two steps (see also [4]).

We first address the problem of detecting whether a set of M polygons monotone in the y-direction is separable in the x-direction. Again we make use of the plane-sweep technique. Let set $S = \{P1, P2, ..., P_M\}$ of M simple n-vertex polygons monotone with respect to the positive y-axis. The following Algorithm 1 determines whether S contains a pair of M simple polygons, monotone with respect to the y-axis. If the polygons do not intersect then it determines a translation order of the polygons.

Algorithm 1: Translating a set of simple monotone polygons.
{*The direction of translation is the positive x-axis*}

Input: A set $S = \{P_1, P_2, ..., P_M\}$ of M simple n-vertex polygons monotone with respect to the positive y-axis.

Output: A translation order of the polygons.

begin
1. Find the vertices of the polygons with the highest and the lowest y-value (denote these vertices as the endpoints of the polygons);
2. Initialize a balanced binary search tree (with null root) and a doubly linked list;
3. For all endpoints of the polygons in descending order do:
   - **if** the endpoint is an upper endpoint of polygon P **then**
     begin
       Insert P into the tree using the procedure
         Insert_Polygon_Into_Tree(Root, P);
       Find the predecessor P' and the successor P'' of P in the tree;
       if P and P' intersect[1] **then** exit program; – *no translation order exists*
       if P and P'' intersect[1] **then** exit program; – *no translation order exists*
       if (P'' exists) **then** insert P as the new predecessor of P'' in the linked list;
       else insert P as the new final element in the linked list;
     end

[1] If P' or P'' do not exist then the polygons do not intersect.

- **if** the endpoint is a lower endpoint of polygon P **then**
  begin
  find the predecessor P' and the successor P'' of P in the tree;
  if P' and P'' intersect[1] **then** exit program; - *no translation order exists*
  else delete P from the tree; (but not from the linked list)
  end

end Algorithm 1

Procedure Insert_Polygon_Into_Tree (Root, P);

{*This procedure inserts a polygon P into a binary balanced search tree.*}

Input:
Root - the root of the tree.
P - a polygon.

begin

while (P not inserted into the tree) **do begin**

1. Find the two vertices, one each of the left and right chains (denoted as v_{Li} and v_{Rj} respectively), of the *Root* such that the y-value of the upper vertex of P (denoted as p) is less then the y-value of v_{Li} (v_{Rj}) and greater than the y-value of v_{Li-1} (v_{Rj+1});
   - *the scan for the new vertices v_{Li} and v_{Rj} starts with the last v_{Lu} and v_{Rw}, of the Root, that have already been found*
2. **if** p is to the right of edge(v_{Li}, v_{Li-1}) and to the left of edge (v_{Rj}, v_{Rj+1}) **then**
   exit program; - *no translation order exists*
3. **elseif** p is to the left of edge(v_{Li}, v_{Li-1}) **then**
   Insert_Polygon_Into_Tree (Left_Son_Of_Root, P);
4. **else** Insert_Polygon_Into_Tree (Right_Son_Of_Root, P); p is to the right of edge (v_{Rj}, v_{Rj+1})

end Procedure

When simple polygons are to be translated in the positive x-direction the visibility hulls of the polygons are found. A translation ordering for S exists if and only if the visibility hulls of the polygons are pairwise non-intersecting. The visibility hulls, which are monotone polygons, are then given as input to Algorithm 1. The output of Algorithm 1 determines the translation order of the simple polygons. If the direction of translation d is not the positive x-axis then the polygons are rotated until the direction of translation is the positive x-direction.

Algorithm 2: Translating a set of simple non-intersecting polygons.

Input:
1) A set $S = \{P_1, P_2, ..., P_M\}$ of M simple non-intersecting n-vertex polygons.
2) The direction of translation d.

Output: A translation order of the polygons.

begin
1. **if** d is not the positive x-axis **then** rotate S such that d is the positive x-axis
2. Find the visibility hulls of the polygons (with respect to the positive x-axis).
3. Let S' be the set of the visibility polygons found in step 2.
   Execute Algorithm 1 on the set S'.

end Algorithm 2

Lemma 3.2 *Algorithm 2 detects whether a set S of M simple n-vertex polygons can be separated in a given direction. If S is separable, the algorithm determines a translation ordering among the polygons. The algorithm runs in $O(Mn + M \log M)$ time and is optimal.*

The algorithm improves on the existing $O(Mn \log Mn)$ algorithm due to Nurmi [3]. It easy to see that the algorithm presented here is optimal since the input size is $O(Mn)$ and sorting reduces to finding a translation ordering a set of M points located on the x-axis. In fact, as shown in [5], this is a lower bound even if we assume that the polygons are stored in memory (thus the time to read the data does not count) and that we are only interested in detecting whether a translation ordering exists (thus the reduction from sorting is no longer valid). Therefore, detecting whether a chain exists is as hard as finding one.

Theorem 3.1 *The existence of a chain in a set S of M n-vertex polygons can be determined in $O(Mn + M \log M)$ time; if at least one chain exists, all chains can be found in the same time. The algorithm is optimal.*

4 Concluding Remarks

In this paper we have discussed and solved a specific geometric problem, that of finding so-called *chains* in a diagram. The problem is of theoretic interest in its own right, while the solution can be applied in dialog systems which use pictorial representations and natural language in the human-computer interface. If a dialog system is to issue a natural language utterance based on or to describe a diagrammatic representation, the diagram must be analysed by an algorithm such as the one presented for finding chains. Based on such analysis, the system can then choose an appropriate wording for the utterance. For example, the algorithm we have presented will identify the switches in Figure 1 as a chain and thus enable the system to describe the highlighted switch to be referred to as "the second switch from the left." Without such geometric processing of the diagram, the system could not use such a formulation.

The specific problem we have discussed is typical for the kind which must be examined to endow knowledge-based systems with flexible processing of diagrams and natural language.

References

[1] A. Barr and E. A. Feigenbaum, editors. *Handbook of Artificial Intelligence.* William Kaufmann, Inc., Los Altos, CA, 1981.

[2] Th. Fehrle, Th. Strothotte, and M. Szardenings. Generating pictorial presentations for advice-giving dialog systems. In M. J. Tauber and P. Gorny, editors, *Visualization in Human-Computer Interaction*, Springer-Verlag, Lecture Notes in Computer Science, Berlin-Heidelberg-New York, 1988. (in print).

[3] O. Nurmi. On translating a set of objects in 2- and 3-dimensional space. *Computer Vision, Graphics and Image Processing*, 36:42–52, 1986.

[4] O. Nurmi and J.-R.Sack. Separating a polyhedron by one translation from a set of obstacles. In *Proc. Workshop on Graph Theory, Amsterdam*, Springer-Verlag, Berlin-Heidelberg-New York, 1988.

[5] D. Nussbaum and J.-R. Sack. *Disassembling composite parts in two and three dimensions via translations.* Technical Report, School of Computer Science, Carleton University, Ottawa, 1988.

[6] T. Ottmann and P. Widmayer. On translating a set of line segments. *Computer Vision, Graphics and Image Processing*, 24:382–389, 1983.

[7] F. P. Preparata and M. I. Shamos. *Computational Geometry: An Introduction.* Springer-Verlag, Berlin-Heidelberg-New York, 1985.

[8] J.-R. Sack and G. Toussaint. Separability of pairs of polygons through single translations. *Robotica*, 5:55–63, 1987.

[9] Th. Strothotte and D. Böcke. Informationsvermittlung in interaktiven wissensbasierten systemen durch bildhafte darstellungen. In S. Maaß and H. Oberquelle, editors, *Software Ergonomie '89*, Teubner, Stuttgart, März 1989. (to appear).

CONVEX BODIES INTEGRAL REPRESENTATIONS

G.Y.Panina *)

The present paper has the following aims:

1. For a translative invariant measure μ on the set E_n of hyperplanes in $\mathbb{R}^n$ to construct so-called wedge density ρ_μ .

2. For convex bodies $K \subset \mathbb{R}^n$ express a measure of the set of hyperplanes intersecting K in terms of ρ_μ .

3. As a conclusion to obtain a decomposition of centrally symmetric convex bodies $K \subset \mathbb{R}^n$ (n is even) in an integral Minkovsky sum:

$$K = \int\limits_{E_n^0} \left(K_\omega^+ - K_\omega^- \right) d\omega,$$

where $K_\omega^\pm$ are the centrally symmetric convex bodies, lying in the hyperplane ω ;

E_n^0 - a set of hyperplanes in $\mathbb{R}^n$, containing O ;

$d\omega$ - a standard Lebesgue measure on E_n^0 .

4. Using this decomposition, to prove the Weile's hypothesis (see [3]) for the case of even dimension of the space.

We shall use the following notations.

$\mathbb{R}^n$ - n -dimensional real space $(n>2)$;

O - origin of coordinates;

E_n - a set of hyperplanes in $\mathbb{R}^n$;

E_n^0 - a set of hyperplanes in $\mathbb{R}^n$, containing O ;

$\mathcal{F}_n$ - a set of flags in $\mathbb{R}^n$, where

a flag f is pair

f (a line L , containing O ; a hyperplane e , containing L);

$\mathcal{F}_n^1$ - a set of superflags in $\mathbb{R}^n$, where

a superflag $\mathcal{F}$ is a pair

$\mathcal{F}$ (a hyperplane ω , containing O ; a $n-1$ -dimensional flag f , lying in the ω);

Ω^{n-1} - an unit sphere in $\mathbb{R}^n$ with a centre in O ;

S_{n-1} - an area of Ω^{n-1}.

1. Wedge density of a measure.

Let μ be a translation invariant (i.e. invariant under translations of $\mathbb{R}^n$) locally finite measure on E_n. We can denote a function ρ_μ on the set $\mathcal{F}_n$ by a formula

*) Fontanka, 27, LOMI,
191011, Leningrad
USSR

$$\rho_\mu(f(L_0,e_0)) = \int_{\mathcal{E}} \frac{\sin^2(e\cap e_0 \hat{,} L_0)}{\cos^{n-3}(e \hat{,} e_0)}\, d\mu(e),$$

where $\mathcal{E}$ is a set of hyperplanes distant from the point O on a distance not more than 1.

ρ_μ is called a wedge density of the measure μ.

2. Calculation of $\mu([K])$.

For a convex body $K \subset \mathbb{R}^n$ let

$$[K] = \{e \in E_n : e \cap K \neq \emptyset\}$$

The following theorems calculate the measure $\mu([K])$ in terms of ρ_μ and parametres of K.

Theorem 1.

Let K be a convex polyhedron.

$\{\nu_i\}_{i=1}^m$ - a set of one-dimensional edges of K;

$|\nu_i|$ - a length of ν_i;

Φ_i - a set of flags $f(L,e)$, where L contains ν_i, and e is a plane of support of K.

df_i - a Lebesgue measure on Φ_i.

Then

$$\mu([K]) = \frac{1}{S_{n-2}} \sum_{i=1}^{m} |\nu_i| \int_{\Phi_i} \rho_\mu(f_i)\, df_i$$

Theorem 2.

Let K be a convex C^2-smooth body.

$S(u)$ - a point on ∂K with a normal $u \in \Omega^{n-1}$;

$R_i(u)$ - an i-th main curvature radius in the point $S(u)$;

$f_i(u)$ - the flag with the line parallel to the i-th main direction in the point $S(u)$, and with the hyperplane with normal u. Then

$$\mu([K]) = \frac{1}{S_{n-2}} \int_{\Omega^{n-1}} \sum_{i=1}^{n-1} R_i(u)\, \rho_\mu(f_i(u))\, du$$

3. Convex bodies integral decomposition.

Let n be even, $n \geq 4$.

It is known the following decomposition of n-dimensional centrally symmetric polyhedron in a Minkovsky's sum of less-dimensional polyhedra (see [1]):

Theorem 3.

Let $K \subset \mathbb{R}^n$ be a centrally symmetric polyhedron.

$\{\Delta_i^j\}$ - a set of it's j-dimentional edges. Then

$$K = \frac{1}{2} \sum_j (-1)^{j+1} \sum_i \Delta_i^j$$

Using this decomposition and the wedge density apparatus, we shall obtain an integral decomposition of smooth n -dimensional bodies.

Let $\omega \in E_n^o$. Every translation invariant measure μ on E_n induces certain translation invariant measure μ_ω on the set E_{n-1} of $n-2$ -dimensional planes lying in ω . Let $\rho_\omega(f)$ be a wedge density of μ_ω . The function $A(\omega, f)$, defined on $\mathcal{F}_n^1$ is called to be a nucleus of a convex centrally symmetric body $K \subset \mathbb{R}^n$, if for any translative invariant measure μ on E_n

$$\mu([K]) = \int_{\mathcal{F}_n^1} A(\omega, f) \rho_\omega(f) \, df \, d\omega$$

Theorem 4.

Let K be a centrally symmetric C^n -smooth body with the support function H . Then K obtaines such a continuous nucleus A , that $A(\omega, f)$ depends only on the value of H and it's n first derivatives in the point $S(\omega)$, where $S(\omega)$ is a point on ∂K which has a normal equal to the normal of ω .

The expression of such a nucleus A in terms of H is known.

A nucleus of a body is determined ambiguous. But a centrally symmetric convex body can be restored by it's nucleus.

Obviously, a sum of bodies's nuclei is a nucleus of the sum. Therefore a problem of integral decomposition of a body in a sum of $n-1$ -dimensional bodies can be reformulated as a problem of decomposition of a nucleus in an integral sum of $n-1$ -dimentional bodies's nuclei.

This problem obtains its solution thanks to the following

Lemma 1.

Let $K \subset \mathbb{R}^n$ be a centrally symmetric convex body, $\omega \in E_n^o$. Then $K \subset \omega$ if and only if K obtains such a nucleus A , that $A(\eta, f) = 0$ for all $\eta \neq \omega$.

Theorem 5. Convex bodies decomposition

Let n be even,

$K \subset \mathbb{R}^n$ - a centrally symmetric convex C^n -smooth body with a support function H .

Then for every hyperplane $\omega \in E_n^o$ there exist such centrally symmetric convex bodies K_ω^+ , K_ω^- , that

a) $K_\omega^+, K_\omega^- \subset \omega$;

b) K_ω^+, K_ω^- depend only on the value of H and it's first n derivatives in the point $S(\omega)$;

c) $K = \int_{E_n^o} (K_\omega^+ - K_\omega^-) \, d\omega$

The least equality is to be understood as an equality of support functions.

4. Weil's hypothesis about zonoid characterisation.

The zonoid characterisation question is of a great interest (see [2]).

Using theorem 5, one can prove the following result for an even dimensional space.

Theorem 6 (Weil's hypothesis, see [3]).

Let n be even, $K \subset \mathbb{R}^n$ - a centrally symmetric convex body with a support function H.

Let for every great sphere

$$\Omega = \Omega^{n-2} \subset \Omega^{n-1}$$

exist a zonoid Z_Ω and a neighbourhood $\mathcal{U}(\Omega)$ such that the boundaries of K and Z_Ω coincide at all points where the normal vector falls in $\mathcal{U}$.

Then K is a zonoid.

The proof is based on the following idea: to obtain a zonoidal decomposition of K (i.e. to express K as an integral sum of segments), one must use the theorem 5, and after that use the zonoidal decompositions of K_ω^+, K_ω^-.

References

[1] Baddeley, A.: Appendix A. In: Ambartzumian R.V. Combinatorial Integral Geometry, John Wiley and Sons Limited Publishers, 1982, 195-211.

[2] Schneider, R., Weil, W.: Zonoids and Related Topics. In: Convexity and its Applications, ed. P.Gruber, J.M.Wills, Basel, 1983, 297-317.

[3] Weil, W.: Blaschkes Problem der lokalen Charakterisierung von Zonoiden. Arch.Math. 29 (1977), 655-659.

Dynamical Systems and Environmental Models

Proceedings of an International Workshop co-sponsored by IIASA and the Academy of Sciences of the GDR held on the Wartburg, Eisenach (GDR), March 17-21, 1986

Edited by Hans Günter Bothe, Werner Ebeling, Alexander B. Kurzhanski, Manfred Peschel

1987. 283 pp – 17 cm x 24 cm –
GDR 35,– M; abroad 38,– DM
ISBN 3-05-500334-9
Order number 763 719 9
Order word: Systems and Models 9044

The book contains the papers presented during the Workshop "Dynamical Systems and Environmental Models" which was organized by the Academy of Sciences of the GDR (Karl Weierstraß Institute for Mathematics) and the International Institute of Applied Systems Analysis. The aim of the workshop was the exchange of ideas between specialists in the theory of dynamical systems and experienced scientists working in the field of environmental models. Accordingly wide is the spectrum of the papers: It reaches from topics in theoretical dynamics which promise to become fruitful in application on the one end over general mathematical models for different kinds of evolutions to concrete dynamical models for environmental processes on the other end.

Yuri M. Svirezhev

Ecological and Demographic Consequences of a Nuclear War

1987. 112 pp – 17 cm x 24 cm – 45 tables
GDR 18,– M; abroad 18,– DM
ISBN 3-05-500193-1
Order number: 763 642 8
Order word: Svirezhev, Nuclear War 6975

The impact of a nuclear war upon ecological and demographic processes is estimated. The direct effect of nuclear detonations, the radioactive contamination of territories, the pollution by heavy metals and acid rain, as well as the effect of the sharp decline in temperature of the atmosphere ("nuclear winter") are taken into consideration. Estimated is the extent of destruction for the ecosystems of different types. The recovery of ecosystems is shown to be extremely slow or impossible. Demographic and genetic predictions for the nuclear war consequences are presented for various regions of the world. The book is intended for specialists in biology and applied mathematics, and for general scientific community as well.

John S. Nicolis

Chaotic Dynamics Applied to Biological Information Processing

1987. 120 pp – 19 figures – 17 cm x 24 cm
GDR 20,– M; abroad 20,– DM
ISBN 3-05-500377-2
Order Number 763 744 7
Order word: Nicolis, Dynamics 9067

Information processing aims at category formation in the (human) cognitive system or the classification of the external (and a good deal of the system's "internal") world in an hierarchy or abstract patterns. These patterns in turn may be used as algorithms for simulating observed physical processes.
In this paper we examine how this abstraction process may be dynamically implemented using the concepts of "Chaotic strange attractors".

Band 1

Inverse and Improperly Posed Problems in Differential Equations

Proceedings of the Conference on Mathematical and Numerical Methods, held in Halle/Saale (GDR) from May 29 to June 2, 1979

Edited by Gottfried Anger
296 Seiten, 1979 (vergriffen)

Band 2

Fundamentals of Computation Theory FCT '79

Proceedings of the Conference on Algebraic, Arithmetic and Categorial Methods in Computation Theory, held in Berlin/Wendisch-Rietz (GDR) September 17–21, 1979

Edited by Lothar Budach
576 Seiten, 1979

Band 3

Decidability and Generalized Quantifiers

By Andreas Baudisch, Detlef Seese, Hans-Peter Tuschik, and Martin Weese
XII, 235 Seiten, 1980

Band 4

Zur Elastizitäts- und Thermoelastizitätstheorie

Potentialtheoretische Methoden in Anwendung auf verschiedene Modelle

Von Lothar Jentsch und Johannes Maul
157 Seiten, 1980

Band 5

Systems Analysis and Simulation 1980

Proceedings of the International Symposium held at Berlin (GDR) September 1–5, 1980

Edited by Achim Sydow
496 Seiten, 1980 (vergriffen)

Band 6

Nichtlineare diskrete Optimierung

von Eberhard Girlich und Michail M. Kowaljow
218 Seiten, 1981

Band 7

Completeness Problems in the Structural Theory of Automata

By Jürgen Dassow
148 Seiten, 1981

Band 8

Darstellung von Lösungen linearer elliptischer Differentialgleichungen

Von Günther Wildenhain
92 Seiten, 1981

Band 9

Introduction to the Spectral Theory of Operators in Spaces with an Indefinite Metric

By I. S. Iohvidov, M. G. Krein, and H. Langer
120 Seiten, 1982 (vergriffen)

Band 10

Nonlinear Time-discrete Systems

A General Approach by Nonlinear Superposition

By Michael Gössel
115 Seiten, 1982 (vergriffen)

Band 11

Preprimal Algebras

By Klaus Denecke
162 Seiten, 1982

Band 12

Ergodic Theory and Related Topics

Proceedings of the Conference held in Vitte/Hiddensee (GDR) October 19–23, 1981

Edited by Horst Michel
232 Seiten, 1982

Band 13

Singularly Perturbed Differential Equations

By H. Goering, A. Felgenhauer, G. Lube, H.-G. Roos, and L. Tobiska
176 Seiten, 1983

Band 14

Leben wir in einer Volterra-Welt?

Ein ökologischer Zugang zur angewandten Systemanalyse

Von Manfred Peschel und Werner Mende
233 Seiten, 1983

Band 15

Markov Decision Problems with Countable State Spaces

Optimality Criteria – Algorithms – Clustering

By Hans Michael Dietz and Volker Nollau
174 Seiten, 1983

Band 16

Stochastische Verfahren zur Suche nach dem Minimum einer Funktion

Von Ryszard Zieliński und Peter Neumann
133 Seiten, 1983

Band 17

Invertibility and Asymptotics of Toeplitz Matrices

By Albrecht Böttcher and Bernd Silbermann
200 Seiten, 1983

Band 18

Structural Induction on Partial Algebras

Introduction to Theory and Application of Partial Algebras – Part II

By Horst Reichel
205 pages, 1984

Band 19

Algebraic Methods for Toeplitz-like Matrices and Operators

By Georg Heinig and Karla Rost
212 pages, 1984

Band 20

Frege Conference 1984

Proceedings of the International Conference held at Schwerin (GDR), September 10–14, 1984
Edited by Gerd Wechsung
408 pages, 1984

Band 21

Numerical Analysis of Selected Semilinear Differential Equations

Edited by Christian Großmann, Thomas Riedrich, and Hartmut Schönheinz
203 pages, 1984

Band 22

Singular Ordinary Differential Operators and Pseudodifferential Equations

By Johannes Elschner
200 pages, 1985

Band 23

Lotka-Volterra-Approach to Cooperation and Competition in Dynamic Systems

Proceedings of the 5th Meeting of UNESCO's Working Group on Systems Theory held on the Wartburg, Eisenach (GDR) at March 5–9, 1984
Edited by Werner Ebeling and Manfred Peschel
271 pages, 1985

Band 24

Convergence Structures 1984

Proceedings of the Conference on Convergence held in Bechyně (Czechoslovakia), September 24–28, 1984
Edited by Josef Novák, Werner Gähler, Horst Herrlich, and Jan Mikusiński
254 pages, 1985

Band 25

Parcella '84

Proceedings of the II. International Workshop on Parallel Processing by Cellular Automata and Arrays held in Berlin (GDR), September 25–27, 1984
Edited by Wolfgang Händler, Tamás Legendi, and Gottfried Wolf
239 pages, 1985 (vergriffen)

Band 26

Multiobjective and Stochastic Optimization Based on Parametric Optimization

By Jürgen Guddat, Francisco Guerra Vasquez, Klaus Tammer, and Klaus Wendler
175 pages, 1985

Band 27/28

Systems Analysis and Simulation 1985
I: Theory and Foundations
II: Applications

Proceedings of the International Symposium held in Berlin (GDR), August 26–31, 1985
Edited by Achim Sydow, Manfred Thoma, and Robert Vichnevetsky
per volume 495 pages, 1985

Band 29

Parcella '86

Proceedings of the III. International Workshop on Parallel Processing by Cellular Automata and Arrays held in Berlin (GDR), September 9–11, 1986
Edited by Tamás Legendi, Dennis Parkinson, Roland Vollmar, and Gottfried Wolf
279 pages, 1986

Band 30

Fuzzy Sets Applications, Methodological Approaches, and Results

Proceedings of the International Workshop on Fuzzy Sets Applications cosponsored by IIASA and convened in cooperation with IAEA held on the Wartburg, Eisenach (GDR), at March 3–8, 1985
Edited by Steffen Bocklisch, Sergei Orlovski, Manfred Peschel, and Yasushi Nishiwaki
233 pages, 1986

Band 31

Mathematical Methods of Specification and Synthesis of Software Systems '85

Proceedings of the International Spring School, held in Wendisch-Rietz (GDR), April 22–26, 1985
Edited by Wolfgang Bibel and Klaus P. Jantke
245 pages, 1986 (vergriffen)

Band 32

A Model Theoretic Oriented Approach to Partial Algebras

Introduction to Theory and Application of Partial Algebras – Part I
By Peter Burmeister
319 pages, 1986

Band 33

Finite Difference Method on Irregular Networks

A Generalized Approach to Elliptic Problems of Second Order
By Bernd Heinrich
206 pages, 1986

Band 34

Modern Algorithms for Large Sparse Eigenvalue Problems

By Arnd Meyer
125 pages, 1987

Band 35

Parametric Optimization and Related Topics

Edited by Jürgen Guddat, Hubertus Th. Jongen, Bernd Kummer, and František Nožička
407 pages, 1987

Band 36

Discretization in Differential Equations and Enclosures

Edited by Ernst Adams, Rainer Ansorge, Christian Großmann, and Hans-Görg Roos
250 pages, 1987

Band 37

Factorization of Measurable Matrix Functions

By Georgi Semenovich Litvinchuk and Ilya Matveyevich Spitkovskii

372 pages, 1987

Band 38

Parallel Algorithms and Architectures

Proceedings of the International Workshop on Parallel Algorithms and Architectures held in Suhl (GDR), May 25–30, 1987

Edited by Andreas Albrecht, Hermann Jung, and Kurt Mehlhorn

205 pages, 1987

Band 39

Parametric Integer Optimization

By Bernd Bank and Reinhard Mandel

138 pages, 1988

Band 40

Computer Analysis of Images and Patterns 1987

Proceedings of the II. International Conference CAIP '87 on Automatic Image Processing, held in Wismar (GDR), September 2–4, 1987

Edited by Leonid P. Jaroslavskiĭ, Azriel Rosenfeld, and Wolfgang Wilhelmi

282 pages, 1987

Band 41

Multivariable Control

A Graph-theoretic Approach

By Kurt J. Reinschke

274 pages, 1988

Band 42

Equilibrium Distributions of Branching Processes

By Alfred Liemant, Klaus Matthes, and Anton Wakolbinger

240 pages, 1988

Band 43

Andreotti-Grauert Theory by Integral Formulas

By Gennadi Markovich Henkin and Jürgen Leiterer

270 pages, 1988

Band 44

Continual Means and Boundary Value Problems in Function Spaces

By Efim Mikhailovich Polishchuk

164 pages, 1988

Band 45

Advances in Mathematical Optimization

Invited papers dedicated to Prof. Dr. Dr. h. c. F. Nožička on occasion of his 70th birthday

Edited by Jürgen Guddat et al.

332 pages, 1988

Band 46/47

Systems Analysis and Simulation 1988

Proceedings of the International Symposium held in Berlin (GDR),

September 12–16, 1988

I: Theory and Foundations
II: Applications

Edited by Achim Sydow, Spyros G. Tzafestas, and Robert Vichnevetsky

426/425 pages, 1988

Band 48

Parcella '88

Proceedings of the IV. International Workshop on Parallel Processing by Cellular Automata and Arrays, held in Berlin (GDR),

October 17–21, 1988

Edited by Gottfried Wolf, Tamás Legendi, and Udo Schendel

379 pages, 1988

Band 49

Algebraic and Logic Programming

Proceedings of an International Workshop held in Gaussig (GDR), November 14–18, 1988
Edited by Jan Grabowski, Pierre Lescanne, and Wolfgang Wechler

278 pages, 1988

Band 50

Asymptotics for Elliptic Mixed Boundary Problems

Pseudo-Differential and Mellin Operators in Spaces with Conormal Singularity

By Stephan Rempel and Bert-Wolfgang Schulze

418 pages, 1989

Band 51

Geobild '89

Proceedings of the 4th Workshop on Geometrical Problems of Image Processing held in Georgenthal (GDR), March 13–17, 1989

Edited by Albrecht Hübler, Werner Nagel, Brian David Ripley, and Günter Werner

198 pages, 1989

The "Geobild" Workshops are designed to stimulate the connections between several fields of mathematics which are applied to Image Processing. The conference topics are: Computational Geometry – Digital/Discrete Geometry – Convex Geometry – Morphology/Texture Analysis – Integral and Stochastic Geometry. In the present volume emphasizes the use of stochastic models and the theory of fuzzy sets, the investigation of higher moments for geometric structures, stereology, axiomatic approaches to discrete geometry, problems of computational geometry and new results for convex bodies.
These fields are not convered in depht, but the collection of papers gives an overview of several directions of research in Image Processing which are to be expected to contribute to a unified mathematical theory of that subject.

Die „Geobild" Symposien sollen Verbindungen fördern zwischen verschiedenen Gebieten der Mathematik, die in der Bildverarbeitung angewendet werden. Die Tagungsthemen sind: Berechenbare Geometrie – Digitale/Diskrete Geometrie – Konvexgeometrie – Morphologie/Texturanalyse – Integral- und Stochastische Geometrie. Zu den Schwerpunkten des vorliegenden Bandes gehören die Benutzung stochastischer Modelle und der Fuzzy-Mengen-Theorie, die Untersuchung höherer Momente für Maßzahlen geometrischer Strukturen, Stereologie, axiomatische Zugänge zur diskreten Geometrie, geometrische Berechnungsprobleme, neue Ergebnisse für konvexe Körper.
Diese Sammlung erfaßt die genannten Gebiete nicht vollständig, gibt aber einen Überblick über verschiedene Forschungsrichtungen in der Bildverarbeitung, von denen erwartet wird, daß sie zu einheitlichen mathematischen Grundlagen beitragen.

ISBN 3-05-500677-1
ISSN 0138-3019

www.ingramcontent.com/pod-product-compliance
Lightning Source LLC
LaVergne TN
LVHW081335011026
849161LV00013BB/2929